Materials Chemistry of Ceramics

Junichi Hojo
Editor

Materials Chemistry of Ceramics

Springer

Editor
Junichi Hojo
Faculty of Engineering
Kyushu University
Fukuoka, Japan

ISBN 978-981-13-9934-3 ISBN 978-981-13-9935-0 (eBook)
https://doi.org/10.1007/978-981-13-9935-0

This Springer imprint is published by the registered company Springer Nature Singapore Pte Ltd.
The registered company address is: 152 Beach Road, #21-01/04 Gateway East, Singapore 189721, Singapore

Preface

Inorganic chemistry is based on the understanding of various elements on the Earth, and the wide variety of functional materials have been created by chemical design starting from atoms or molecules. They are extensively used as essential and crucial matters in our life and industry activities and have also contributed to the development of new fields of environment, energy, informatics, biomaterials, and so on. Ceramics are typical inorganic materials. They have made a widespread evolution by generation of new substances, innovation of new technologies, and precise design of structure and morphology. The concept of material chemistry is a key for production and utilization of practical ceramics. This book introduces a comprehensive content from fundament to application on the backbone of inorganic synthetic chemistry and solid-state chemistry.

Ceramic industries have advanced from traditional ceramic manufactures: pottery and porcelain, refractory, glass, and cement. They are produced by firing process using kiln. The raw materials are natural silicate compounds. Silicates are composed of silica, alumina, alkaline oxide, and alkaline earth oxide and easily mined on the Earth's crust. The advantage is the low melting point. Therefore, sintered ceramics and melted glasses do not require a very high temperature. Meanwhile, new ceramics have developed using artificial raw materials during the past half-century. Industrial new ceramics seem to start from electric insulator. Ceramic insulators are superior in mechanical strength and insulation performance to suspend high-voltage electrical power lines. New ceramics mean the usage of new substances and compounds. Oxides are typical ingredients of ceramics, but highly pure oxides are required. Non-oxides such as nitride, carbide, and boride have high strength and thermal stability but are naturally not afforded. Various inorganic compounds can be made by a combination of numerous elements in binary, ternary, and multiple systems, leading to appearance of a variety of electromagnetic devices. Fine ceramics aim at new and high function based on material design, that is, homogeneous composition, fine microstructure, and morphological control such as nanoparticle, nanosheet, nanofiber, etc. Such a trend of new materials has induced high-pressure and high-temperature technologies, or new synthetic methods like chemical vapor deposition

(CVD), sol-gel technique, etc. As a result, many kinds of ceramics have been practically applied as high-temperature structural material, mechanical part, cutting tool, semiconductor, dielectric, magnetic device, optical fiber, artificial bone, and so on.

Ceramic structures are classified into single crystal, polycrystal, and amorphous crystal/glass. Single crystal has a regular arrangement of atoms in whole body. Amorphous crystal and glass have no long-range regular arrangement with random atomic bonding. Polycrystal is an aggregate of small crystalline grains. Since most of ceramics are fabricated by sintering of raw powder, they are obtained in polycrystalline form. Even by homogeneous processes (CVD, sol-gel), the final products are mostly polycrystalline. Chemical design means the synthesis from atomic or molecular level through chemical route and the intentional control of microstructure and morphology toward desirable properties. Thus, we need the basic knowledge in chemical synthesis, crystal structure, and physical and chemical characteristics of solids to establish the practical technologies for functional material design. This book describes the structure and synthesis of ceramics in the anterior half and then the practical functions of ceramics in the posterior half. The last organic-inorganic hybrid is a new scope beyond the category of ceramics. It is our great pleasure if all of the readers understand what ceramics are and how we can use it. Please create the future ceramics for the next generation.

Fukuoka, Japan
November, 2019

Junichi Hojo

Contents

Contributors

Junichi Hojo Faculty of Engineering, Kyushu University, Fukuoka, Japan

Miki Inada Center of Advanced Instrumental Analysis, Kyushu University, Fukuoka, Japan

Masakazu Kawashita Department of Inorganic Biomaterials, Institute of Biomaterials and Bioengineering, Tokyo Medical and Dental University, Tokyo, Japan

Akira Kishimoto Graduate School of Natural Science and Technology, Okayama University, Okayama, Japan

Krisana Kobwittaya Department of Science and Advanced Technology, Graduate School of Science and Engineering, Saga University, Saga, Japan

Junichi Matsushita Department of Materials Science, Tokai University, Hiratsuka, Japan

Tsugio Sato Institute of Multidisciplinary Research for Advanced Materials, Tohoku University, Sendai, Japan

Yoshiyuki Sugahara Department of Applied Chemistry, School of Advanced Science and Engineering, Waseda University, Tokyo, Japan
Kagami Memorial Research Institute for Materials Science and Technology, Waseda University, Tokyo, Japan

Satoshi Tanaka Department of Materials Science and Technology, Nagaoka University of Technology, Nagaoka, Japan

Junichi Tatami Graduate School of Environment and Information Sciences, Yokohama National University, Yokohama, Japan

Takanori Watari Department of Science and Advanced Technology, Graduate School of Science and Engineering, Saga University, Saga, Japan

Shu Yin Institute of Multidisciplinary Research for Advanced Materials, Tohoku University, Sendai, Japan

Toshinobu Yogo Institute of Materials and Systems for Sustainability, Nagoya University, Nagoya, Japan

Chapter 1
Crystal Structures of Inorganic Materials

Junichi Hojo and Miki Inada

Abstract Ceramics have high hardness and mechanical strength, excellent chemical and thermal stabilities, and particular electromagnetic and optical properties. Based on these unique properties, they have been widely used in our life, industry activities, and various environmental and energy fields. The durability of ceramics is derived from strong chemical bonding, that is, covalent and ionic bonds, and tight atomic packing. The electrical properties are dominated by crystal and electronic band structures and typically classified into insulator, dielectric, semiconductor, and metallic conductor. Typical covalent crystals are diamond, graphite, and hexagonal boron nitride (h-BN). Diamond, having 3D carbon network, is extremely hard, whereas graphite and h-BN are soft due to the hexagonal layer structure. Structural principle in ionic crystal is the alternate arrangement of cation and anion different in charge and size. The coordination configuration consisting of centered cation and surrounding anions is the elemental unit: for example, silica is constructed by SiO_4 tetrahedrons and titania by TiO_6 octahedrons. Lattice defects, such as vacancy and interstitial atom, provide a large influence on semiconductor and ionic conductor. Glass has a random network structure and has been applied in optical fields. Recent trend is the dimensional control of morphology from zero- to three-order. New functions of ceramics have been created using nanoparticle, nanofiber, nanosheet, hybrid, and nanoporous materials.

Keywords Chemical bonding · Crystal · Glass · Defect structure · Dimensional control

J. Hojo (✉)
Faculty of Engineering, Kyushu University, Fukuoka, Japan
e-mail: hojo.junichi.090@gmail.com

M. Inada
Center of Advanced Instrumental Analysis, Kyushu University, Fukuoka, Japan
e-mail: inada.miki.300@m.kyushu-u.ac.jp

J. Hojo (ed.), *Materials Chemistry of Ceramics*,
https://doi.org/10.1007/978-981-13-9935-0_1

1.1 Fundamental Structures and Properties of Materials

Materials are generally classified into ceramic, metallic, and organic ones. Organic materials are comprised mainly of carbon, hydrogen, oxygen, and nitrogen. Although the number of constituent element is small, there are many organic compounds owing to the variety of molecular structures based on directional covalent bonding. Plastics are formed by connection of small monomers. The main chain is strong due to the covalent C-C bond, but hydrogen bonding or van der Waals bonding between main chains is weak. Therefore, they are soft and flexible, but inferior in heat resistance. Typical metallic materials are iron, aluminum, copper, etc. Most of metallic elements belong to transition metals. Metal crystals are constructed by regular packing of spherical atoms. Metallic bonding is a kind of covalent bonding in a broad meaning. The heat resistance is high compared to organic compounds. Since electrons are not localized but extensively shared among atoms, electrons easily move, causing high electrical and thermal conduction. The displacement of atoms is also easy, causing high ductility. Ceramics are mainly composed of positive and negative elements in the left and right sides, respectively, of the periodic table. They are connected by covalent and ionic bonding. Since shared electrons and electrostatic attraction among atoms cause strong bonding force, they are mechanically rigid and undeformable, and thermally and chemically stable. Furthermore, the electronic state is wide-ranged owing to the combination of various atoms, leading to the appearance of versatile electronic materials, such as insulator, dielectrics, semiconductor, and so on.

Typical fabrication process and schematic model of microstructure in ceramics are demonstrated in Fig. 1.1. Most of ceramics are fabricated by powder forming and subsequent high-temperature sintering, resulting in polycrystalline structure composed of agglomerated crystal grains. There are grain boundaries between grains with different crystal orientations. If raw powders include metallic impurities, foreign particles remain as inclusions. If the sintering is not completed, pores remain

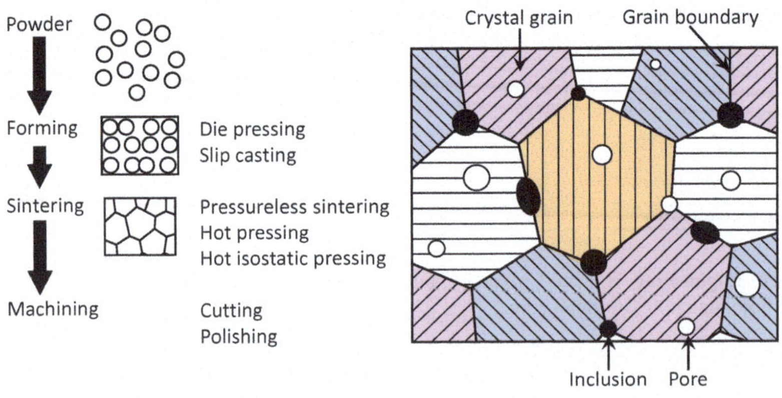

Fig. 1.1 Schematic of fabrication process of ceramics and model of sintered texture

in sintered body. Grain boundary, inclusion, and pore seriously affect the mechanical strength. The total structure of ceramics is classified into texture and transgranular structure. The texture means the agglomerated-grain structure. The final properties of ceramics are affected by various factors: grain size, pore size, volume, and so on. The transgranular structure is involved with atomic bonding and crystal structure. In this chapter, the crystal structures of ceramics are introduced for better understanding of the principal properties and functions.

1.2 Crystal Structure of Inorganic Material

1.2.1 Packing Manner of Rigid Spheres

Crystal structures are simply explained from packing manner of rigid spheres. Figure 1.2 demonstrates fundamental crystal structures. The minimum unit of crystal is called unit lattice or unit cell. Simple cubic (SC) structure has four atoms in the corners. Body-centered cubic (BCC) structure includes an additional atom in the body center. Close packing structures are most stable. As shown in the packing

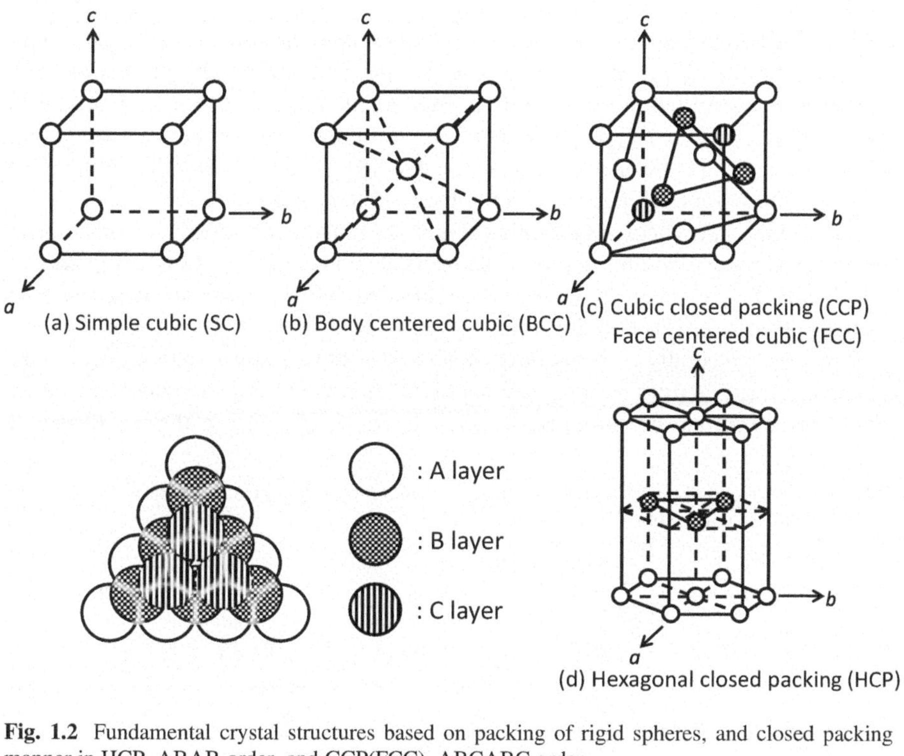

Fig. 1.2 Fundamental crystal structures based on packing of rigid spheres, and closed packing manner in HCP, ABAB order, and CCP(FCC), ABCABC order

manner of spheres, the stable packing is based on triangular arrangement. When putting spheres on the first layer (A), they are settled on the concave of three spheres in underlayer to form the second layer (B). There are two possibilities for the third layer. When spheres are placed on the same positions of (A), they form hexagonal close packing (HCP) structure in the ABAB order. It is normally illustrated as hexagonal column, but the unit cell is rhombohedral column with equivalent a- and b-axes and c-axis. When the third spheres are placed on the different positions from (A), the new layer is recognized as (C), forming cubic close packing (CCP) structure in the ABCABC order. The unit cell is illustrated as face-centered cubic (FCC) structure placing atoms in the center of cube face. The close packing direction is in the diagonal of cube.

SC structure is scarcely observed because the open space inside the cube makes it unstable. Metals have simple crystal structures of BCC, CCP, and HCP because spherical atoms with the same size are packed. In contrast, ceramics have different types of complicated crystal structures because constituent atoms are different in size and charge.

1.2.2 Coordination Configuration

In crystal structures, atoms are surrounded by the other atoms. It is called coordination configuration which is determined by atomic size and bonding direction. The variation of coordination configuration with ionic radius and bonding orbital is explained as in Table 1.1. Typical coordination figures and numbers are triangular (3), tetrahedral (4), octahedral (6) and cubic (8). The coordination number of 12 means the close packing, but it is not the case in ionic crystals.

Coordination polyhedrons of ionic crystals are illustrated in Fig. 1.3. Ionic bonds form between positive and negative elements. For instance, MgO is composed of Mg^{2+} and O^{2-}. Cations and anions are bonded by electrostatic attraction (the Coulomb's force). The bonding force depends on distance between ions, but not on direction. Generally, cations have smaller size than anions. Cations form with release of electrons from atoms, resulting in the increase of effective nuclear charge. The outer electrons are strongly attracted by nuclear, causing the shrinkage of

Table 1.1 Variation of coordination configuration with radius ratio of cation to anion and shape of hybrid orbital of center atom

	Cation/anion radius ratio			
Coordination number	Theoretical	Range	Configuration	Hybrid orbital
3	0.155	0.155–0.225	Triangular	sp^2
4	0.225	0.225–0.414	Tetrahedral	sp^3
6	0.414	0.414–0.732	Octahedral	d^2sp^3
8	0.732	0.732–1.000	Cubic	
12	1.000	Closed packing		

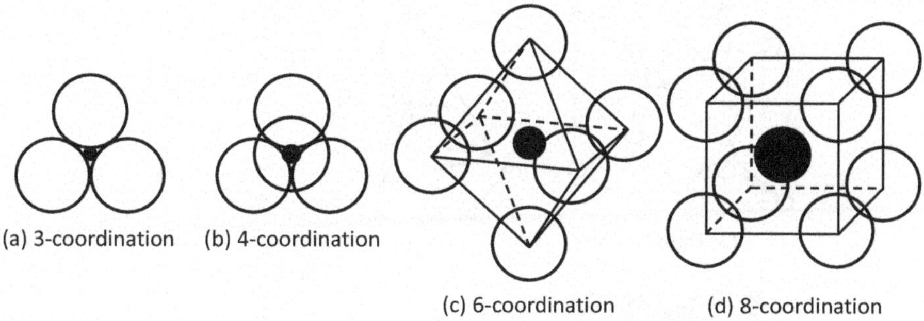

Fig. 1.3 Typical coordination configuration in ionic crystals. Small cation is centered and surrounded by large anions

cations. On the other hand, anions form by addition of extra electrons, which lowers the effective nuclear charge and causes the expansion of anions. In ionic crystals, small cations are settled in voids of packing structure of large anions. The coordination number of anions surrounding cation is determined by the ionic radius ratio. Large cation can be surrounded by many anions. Therefore, the coordination number increases with an increase in radius ratio of cation to anion. It is called "Pauling's law." Theoretical ratio means the calculated one when cation just fits the void of surrounding anions. When the cation-to-anion radius ratio is equal to the theoretical one or above, the crystal structure is stable. When the cation size is smaller than the theoretical one, the structure becomes unstable because small cation cannot be settled in the center of interstice surrounded by anions.

Covalent bonds form by sharing of localized electrons in overlapping of atomic orbitals, in which the orbital direction determines the coordination configuration. Generally, it can be explained from hybrid orbitals of centered cation which are constructed by s, p, and d orbits: sp^2 hybrid (3-coordination, triangular), sp^3 hybrid (4-coordination, tetrahedron), and d^2sp^3 (6-coordination, octahedron). These coordination figures are similar to those in ionic bonding. The compounds constructed by different elements exhibit both covalent and ionic behaviors. When the coordination configuration expected from ionic size is the same as in atomic orbital, the crystal structure is stable. SiO_2 is expected to have 4-oxygen coordination from both ionic size and atomic orbital of silicon. It is hard and thermally stable. ZnO is expected to have 6-oxygen coordination from ionic size but 4-oxygen coordination from atomic orbital of zinc. Therefore, ZnO is less stable than SiO_2.

1.2.3 *Covalent Crystals*

Typical covalent material is a carbon family with various crystal structures, including diamond, graphite, fullerene, carbon nanotube, and graphene. The crystal structures and bonding orbitals of diamond and graphite are illustrated in Fig. 1.4.

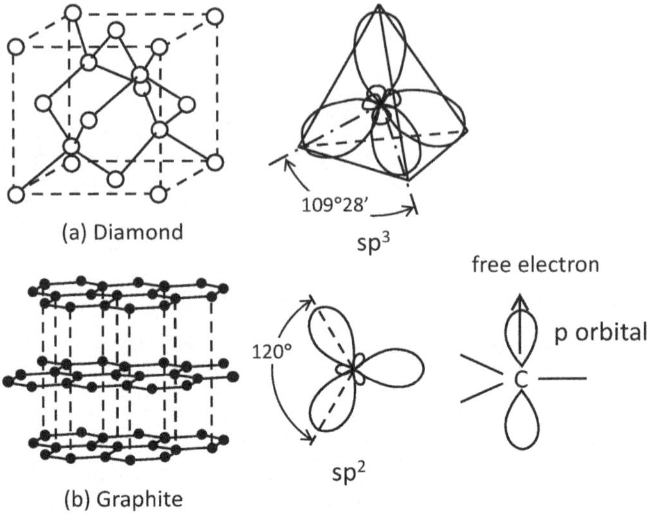

Fig. 1.4 Crystal structures of diamond derived from sp^3 hybrid orbital and graphite derived from sp^2 hybrid orbital. There is a free electron in p orbital perpendicular to sp^2 hybrid orbital of graphite

Fig. 1.5 Crystal structure of h-BN and electron configuration in p orbitals perpendicular to sp^2 hybrid orbitals of boron and nitrogen

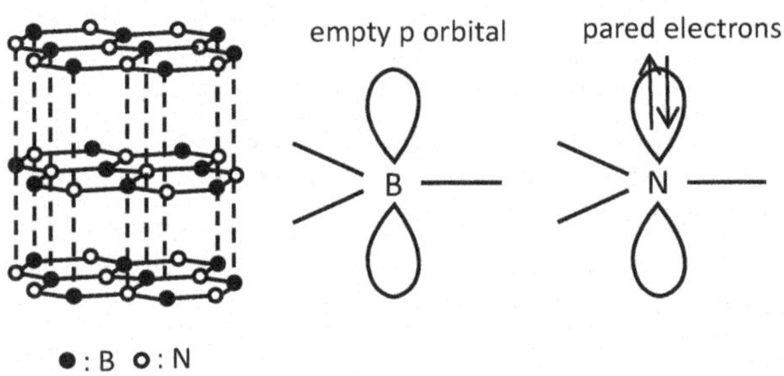

Diamond has three-dimensional network structure consisting of carbon tetrahedrons with 4-coordination. This structure is derived from sp^3 hybrid orbital. Graphite has two-dimensional layer structure consisting of stacked hexagonal planes with 3-coordination. This structure is derived from sp^2 hybrid orbitals. Diamond has an extremely high hardness because all four orbitals covalently bond. Since there is no free electron, it is an insulator and the single crystal is transparent. Graphite has strong covalent bonds in the in-plane direction of hexagonal network, but the van der Waals bonding between planes is so weak that hexagonal planes easily exfoliate. Since graphite is soft, it has been applied as solid lubricant. One p orbital remains in the perpendicular direction of sp^2 hybrid orbital. One electron exists in this p orbital, and the unpaired electron works as free electron. Therefore, graphite has high electrical conductivity due to the easy electron transport along hexagonal planes. The free electrons absorb light in wide range, and, therefore, graphite looks black.

Boron nitride forms graphite-like hexagonal structure (Fig. 1.5). Boron and nitrogen atoms are alternately bonded and form BN hexagonal planes through covalent bonding between sp^2 hybrid orbitals of boron and nitrogen. Since the bonding between planes is weak, the hexagonal boron nitride (h-BN) is soft. The p orbital in the perpendicular direction of sp^2 hybrid orbitals is empty in boron, but has two electrons with antiparallel spins in nitrogen. Since the pared electrons are stable and localized, they cannot move. Therefore, h-BN is an insulator. Since light is scattered without absorption, the powder looks white. The h-BN is called "white graphite." Boron nitride can form cubic crystal structure similar to diamond under high pressure. The cubic boron nitride (c-BN) also has high hardness. Silicon carbide (SiC) has diamond-like crystal structure, and the hardness is high. The similarity in crystal structure among carbon, boron nitride, and silicon carbide is easily understood in the periodical table. Boron and nitrogen are placed in the left and right from carbon, and silicon is placed just under carbon. They can form the similar covalent bonding using s and p orbitals.

1.2.4 Ionic Crystals

Ionic crystals are formed by alternate arrangement of cation and anion. The crystal structures are determined by coordination number based on ratio of ionic radius and chemical composition. Multicomponent ceramics have more complex crystal structures.

1.2.4.1 Coordination Number and Crystal Structure

When the size of cation is large, the cation can be surrounded by many anions. That is, the crystal structures with high coordination number become stable. This is demonstrated for one-to-one compounds as follows.

Zinc blende and wurtzite structures have 4-coordination number. Zinc blende is the mineral name of β-ZnS. β-SiC belongs to this structure. Zinc blende structure is illustrated in Fig. 1.6. Anions form CCP structure. Cation is surrounded by 4 anions in the shape of tetrahedron. The number of 4-coordination position is 8 in the unit cell. Four cations are placed in half of the 4-coordination positions in the sterically alternate manner to reduce the repulsion between cations. Coordinate graphic is convenient to indicate the positions of atoms. In zinc blende structure, anions are located at the corners and faces of cube. The origin of unit cell is situated at the basal plane. The positons of origin, half unit, and lattice unit are represented by 0, 1/2, and 1, respectively. Cations are located at 1/4 and 3/4 units from origin. Wurtzite is the mineral name of α-ZnS. Anions form HCP structure and cations are placed in half of the 4-coordination positions. This structure is easy to understand as hexagonal column (Fig. 1.7a). However, the unit cell should be written as rhombohedral column (Fig. 1.7b). α-SiC, AlN, and ZnO have this structure. AlN has been applied as insulating substrate and ZnO as semiconductor.

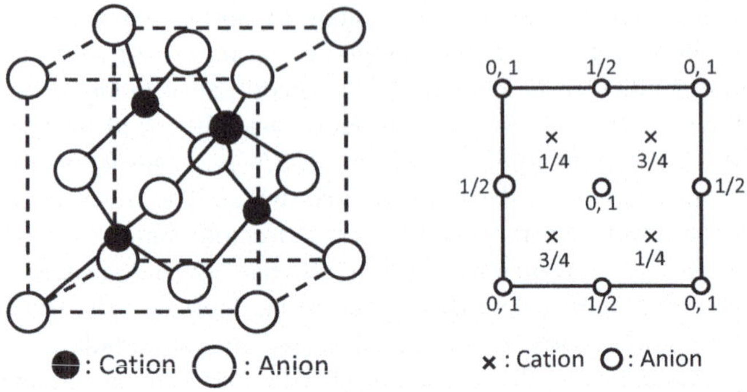

Fig. 1.6 Zinc blende structure and coordinate graphic (projection figure)

Fig 1.7 Wurtzite structure illustrated in two manners: hexagonal and rhombohedral columns

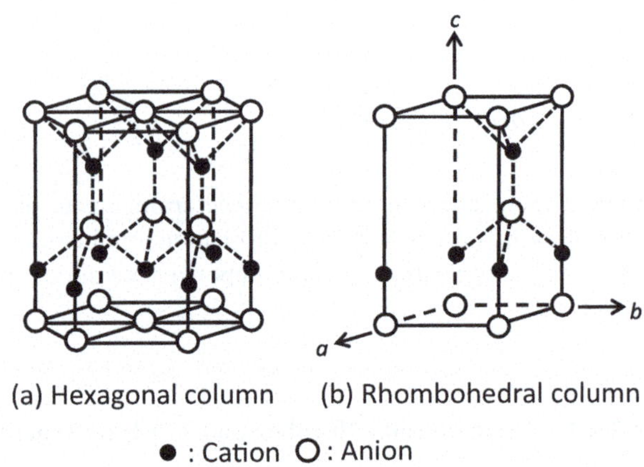

Fig. 1.8 Crystal structures of rock salt and CsCl types

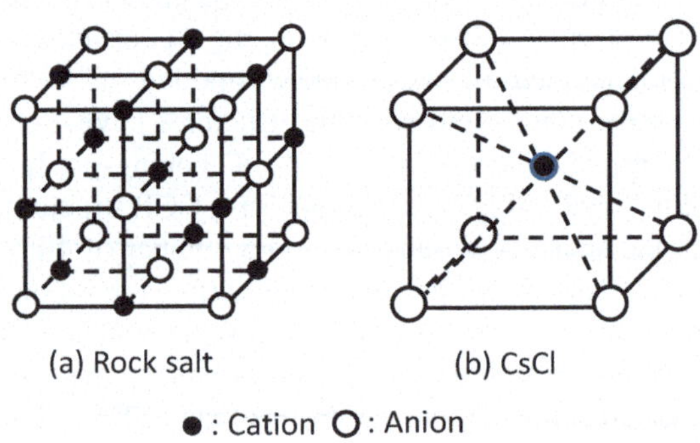

Typical crystal structure with 6-coordination number is rock salt type. Rock salt is the mineral name of NaCl. The crystal structure is illustrated in Fig. 1.8a. Anions form CCP structure. Cation is surrounded by 6 anions in the shape of octahedron. All 6-coordination positions are occupied by cations. Cation and anion, respectively, form FCC structure. However, this structure cannot be assigned to FCC, because the crystal consists of different ions. The rock salt structure is observed in many metal compounds, such as halide (LiF, NaCl, KCl, AgCl), oxide (MgO, CaO, MnO, CoO, NiO), nitride, carbide (TiN, TiC), etc.

CsCl structure has 8-coordination number as seen in Fig. 1.8b. Anions are located at the corners of simple cube, and the body center is occupied by cation. However, this structure should not be assigned to BCC. NH_4Cl has this structure. Although atomic radius of nitrogen is small, cation size of NH^{4+} becomes large. As described here, the large cation can form high-coordination crystal structures.

1.2.4.2 Various Chemical Compositions

Metal ions have various electric charges. Depending on the electric charge, the cation-to-anion ratio is changed. It makes the crystal structure complicated. As one-to-two compound, SiO_2 and TiO_2 are typically cited. SiO_2 is considered as ionic crystal, but the contribution of covalent bonding is large between silicon and oxygen. Therefore, the crystal structure is explained based on SiO_4 tetrahedron. SiO_4 units are shared at the corner of tetrahedron by covalent bonding through oxygen. The bonding angle at the corner is flexible. Therefore, there are many polymorphic structures: α and β phases of quartz, tridymite, and cristobalite.

Rutile is one of the crystal structures of TiO_2 (Fig. 1.9). Anions form HCP structure, but it is distorted to form tetragonal unit cell. Cation is placed in the 6-coordination position. This structure is also explained in the connection manner of

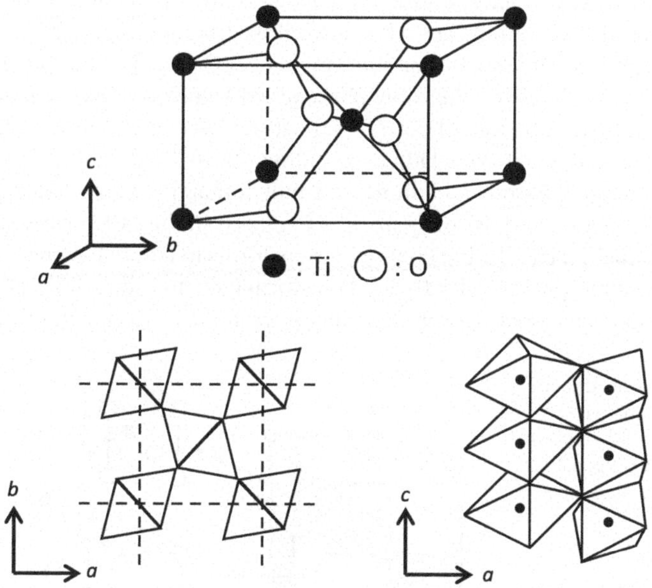

Fig. 1.9 Crystal structure of rutile: unit cell and connection manner of TiO_6 octahedrons

TiO_6 octahedrons. As seen in Fig. 1.9, TiO_6 units are shared at the edge of octahedron in c-axis direction, and the connected chains are bonded by corner sharing in a- and b-axes. The other crystal structures of TiO_2 are anatase and brookite. They are also explained in different connection manners of TiO_6 octahedron. The connection model of oxygen polyhedron is convenient to explain the complicated crystal structure. TiO_2 has been applied as white pigment, photocatalyst, semiconductor, dielectric, etc. SnO_2 has the rutile structure. It has been used as transparent semiconductor, gas sensor, and so on.

Corundum is the mineral name of α-Al_2O_3. Anions form HCP structure. Cations are placed in 2/3 of the 6-coordination positions. The distribution of cations is not random, but periodical from A to B layers in HCP structure. The crystal structure of α-Al_2O_3 is also explained by the connection model of AlO_6 octahedron, in which AlO_6 units are shared at the edge and face of octahedron. The stability of sharing of oxygen polyhedron decreases in the sequence of corner, edge, and face because the short distance between cations increases electrostatic repulsion. Therefore, Al_2O_3 tends to form the other polymorphic structures such as γ-Al_2O_3 with spinel structure. Among them, α-Al_2O_3 is most stable. It has been widely applied as mechanical part, cutting tool, crucible, sodium vapor lamp, IC substrate, etc., because of the balanced excellent properties in hardness, heat and chemical resistance, electrical insulation, heat conduction, and optical transparency.

1.2.4.3 Solid Solution and Complex Compound

ZrO_2 has a cubic type of fluorite structure with 8-coordination number. Fluorite is the mineral name of CaF_2. ThO_2 and UO_2 have this structure. Since the cation size is large, they can have high-coordination number. ZrO_2 has been applied as typical ceramics material. However, the cubic phase (c-ZrO_2) needs to be stabilized by solid solution. The crystal structure of c-ZrO_2 is illustrated as overall view in Fig. 1.10. SC oxygen lattice and CsCl-type lattice including Zr in 8-oxygen coordination are alternately arranged. Figure 1.11 shows the unit cell and the coordinate graphic. Zr forms FCC structure, and SC oxygen cube is placed in the FCC center. c-ZrO_2 is a high-temperature phase. During cooling, it transforms to tetragonal phase at 2370 °C and to monoclinic phase at 1000 °C. Transformation to monoclinic phase causes abnormal large-volume expansion, resulting in collapse of ZrO_2 ceramics. Since the

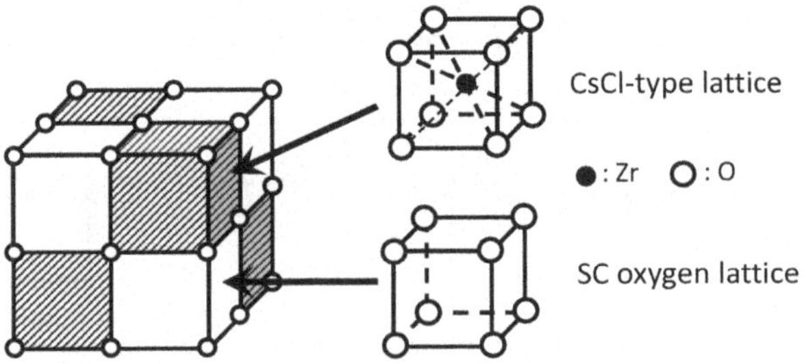

Fig. 1.10 Crystal structure of c-ZrO_2: alternative arrangement of SC oxygen lattice and CsCl-type lattice including Zr at the center

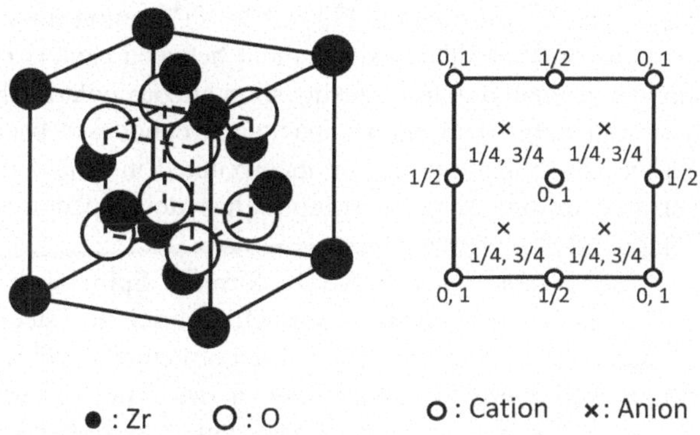

Fig. 1.11 Crystal structure of c-ZrO_2: unit cell and coordinate graphic

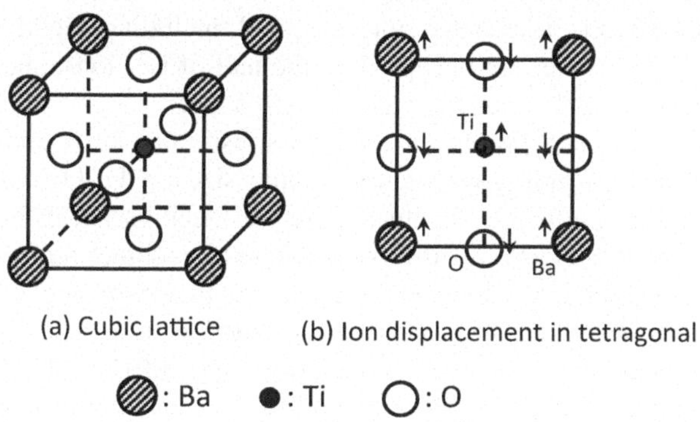

Fig. 1.12 Crystal structure of cubic perovskite ($BaTiO_3$) and displacement of ions in tetragonal phase

ionic radius of Zr^{4+} (0.082 nm) is smaller than optimal radius (0.102 nm) in 8-oxygen coordination, c-ZrO_2 becomes unstable at low temperatures. The c-ZrO_2 can be stabilized by substitutional solution of Ca^{2+} (0.103 nm) and Y^{3+} (0.096 nm) having larger ionic radius. This is the "stabilized zirconia." The substitution of low-valent cation for Zr^{4+} forms oxygen vacancies, leading to the application to oxide ion conductor. Partially stabilized zirconia (PSZ) is produced by reducing amount of stabilizer. The tetragonal zirconia polycrystal (TZP) has high toughness, leading to mechanical and structural application.

Complex oxides include different metal ions at the cation site. Typical one is perovskite represented in ABX_3 formula. Perovskite is the mineral name of $CaTiO_3$. $BaTiO_3$, $SrTiO_3$, $PbTiO_3$, and $PbZrO_3$ have this structure and have been applied as dielectric and piezoelectric materials. A-site cation is large and B-site cation is small. The crystal structure of $BaTiO_3$ is explained here. Large Ba^{2+} cation is coordinated by 12 oxygens and small Ti^{4+} cation by 6 oxygens. The structure based on TiO_6 octahedron is generally used as seen in Fig. 1.12a. Ba^{2+} cations are placed at the corners and TiO_6 unit inside the unit cell. The ideal structure is cubic, and the cubic phase is stable above 120 °C. It is distorted around room temperature and changes to tetragonal phase. The reason is the small size of Ti^{4+} cation compared to the optimal size in 6-oxygen coordination. For practical application, the tetragonal phase is important. As illustrated in Fig. 1.12b, Ti^{4+} cation somewhat sifts from the center of O^{2-} coordination. The displacement between centers of plus and minus charges is aligned in small domain, causing spontaneous polarization. Thus, tetragonal $BaTiO_3$ has been applied as high capacitance condenser because of the ferroelectricity. Various metal cations can be incorporated in A and B sites of perovskite. Such a versatile design of solid solution has developed many functions of perovskite compounds.

Spinel is represented in AB_2X_4 formula. Spinel is the mineral name of $MgAl_2O_4$. This is main component of refractory brick for steel making, and also used as transparent ceramics etc. The crystal structure is cubic, and explained like fluorite structure in alternate arrangement of two types of cubic units including A-site or B-site. A-site has 4-oxygen coordination and B-site has 6-oxygen coordination. When A-sites are occupied by A^{2+} ions and B-sites by B^{3+} ions, that is, in $(A)[B]_2O_4$, it is called normal spinel, including $MgAl_2O_4$, $ZnFe_2O_4$, $CdFe_2O_4$, etc. When A-sites are occupied by the half of B^{3+} ions, and B-sites by A^{2+} ions and the half of B^{3+} ions, that is, in $(B)[A,B]O_4$, it is called inverse spinel, including $MgFe_2O_4$, $NiFe_2O_4$, etc. Such a substitution in A and B-sites is possible between A and B atoms with similar ionic sizes. γ-Fe_2O_3 (maghemite) is represented as $(Fe)[Fe_{5/3}\square_{1/3}]O_4$, in which there are cation vacancies in B-sites. Magnetic iron oxide compounds are called ferrites. Generally, spinel ferrites are easy for magnetization and inversion of magnetic dipole, and, therefore, applied as soft magnetic material in magnetic recording and high frequency transformer. Fe_3O_4 (magnetite) has the mixed valence in the inverse spinel type of $(Fe^{3+})[Fe^{2+},Fe^{3+}]O_4$. This is classified into hard magnetic material and used as permanent magnet.

1.3 Defect Structures in Crystal

Crystal structures are understood on the assumption of perfect ones, in which atoms are ideally arranged in an orderly manner at lattice points. However, there are lattice defects in real crystals. This is caused by thermal oscillation of atoms. Unless the absolute temperature is 0 K, atoms jump out of lattice points with a certain probability and move to the other positions. Defect structures are thermodynamically considered to be stable. The Gibb's free energy increases in the term of enthalpy due to decrease in the number of atomic bonds, but decreases in the term of entropy due to the structural diverseness of atomic arrangement, totally leading to the decrease in free energy by defect formation. There are some kinds of lattice defects. Point defects form at the lattice points themselves. Extended defects form by three-dimensional aggregation and rearrangement of point defects: cluster, sheared structure, and block structure. Dislocation is the line defect by linear arrangement of point defects, and plays an important role in plastic deformation of metal crystals. Point defects provide a large influence on atomic diffusion, and electromagnetic and optical properties of ceramics.

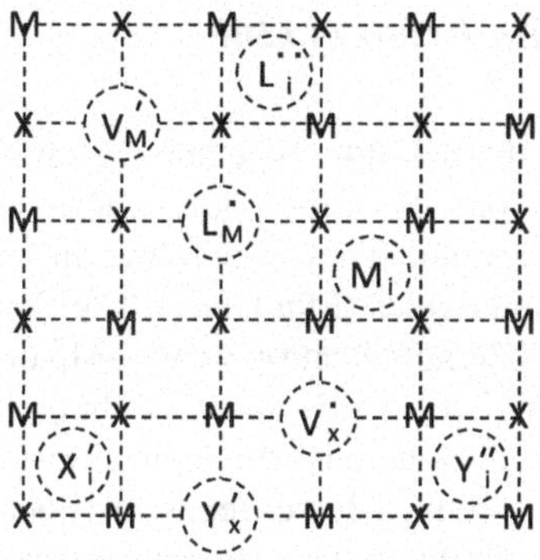

Fig. 1.13 Point defects in $M^{+}X^{-}$ crystal. Foreign cation (L^{2+}) and anion (Y^{2-}) are included as impurities

1.3.1 Point Defects

Fundamental point defects in solid crystal are vacancy and interstitial atom. Impurity atom and associated center are also considered. In many cases, different point defects coexist in crystal structure. The point defects are summarized as follows:

Vacancy: regular lattice points are empty.
Interstitial atom: some atoms exist between lattice points.
Impurity atom: foreign atoms occupy lattice point or interstitial site.
Associated center: Point defects interact with each other by the Coulombic force.

Point defects form in covalent and ionic crystals. The defect structures are determined by atomic size and ionic charge balance. Figure 1.13 illustrates the schematic model of point defects in ionic crystal. These point defects are notated in a peculiar manner to clarify defect reaction among ions and electrons. The Kröger-Vink notation is widely used to symbolize point defects as A_a^b. "A" means atomic symbol. Vacancy is represented by "V." The subscript "a" means atomic position on lattice which is denoted by original atom. Interstitial site is represented by "i." The superscript "b" means effective charge, indicating the relative charge based on original charge of lattice point. It is denoted as plus ($^{\bullet}$), minus ($'$), and neutral (^{x}). This notation is explained for $M^{+}X^{-}$ crystal as below. Vacancies at M and X-sites are denoted as V_M' and $V_X^{\bullet}$, respectively. This means that vacancies exist at M and X-sites. The real charge of vacancy should be neutral. Based on the charge of M^{+} and X^{-}, the effective charge of vacancy negatively sifts by minus one for M-site, and positively by plus one for X-site. M^{+} and X^{-} at interstitial sites are denoted as $M_i^{\bullet}$ and X_i', respectively. This means that M^{+} and X^{-} exist at i-sites. Since the real charge of interstitial site is neutral, the effective charges of M^{+} and X^{-} at interstitial sites are plus one and minus one, respectively. Impurity cation, L^{2+}, substituted for M^{+} is denoted as $L_M^{\bullet}$, and impurity anion, Y^{2-} , substituted for X^{-} as Y_X'. L^{2+} and Y^{2-} at interstitial sites are denoted as $L_i^{\bullet\bullet}$ and Y_i'', respectively.

1.3.2 The Schottky Defect and the Frenkel One

Typical point defects in ionic crystals are the Schottky type and the Frenkel type as shown in Fig. 1.14. The Schottky defect consists of the paired vacancies at cation and anion sites as V_M'-$V_X^{\bullet}$ in M^+X^- crystal. They are found in many metal compounds, such as halides (NaCl, etc.) and oxides (MgO, etc.). The Frenkel defect consists of the pair of cation vacancy and interstitial cation as V_M'-$M_i^{\bullet}$ or the pair of anion vacancy and interstitial anion as $V_X^{\bullet}$-X_i'. CaF_2 and UO_2 have the Frenkel defects because the interstitial sites are large. Generally, the ionic radius of cation is small in ionic crystals, and therefore, the Frenkel defect tends to appear in cation sublattice. AgBr and AgCl have the Frenkel defects. This is explained as polarizable Ag ions easily enter interstitial sites. Table 1.2 demonstrates some reaction formulas for formation of the Schottky defects and the Frenkel ones. The Schottky defect is written to form from "null." The Frenkel defect is written to form from neutral atom. Different defect structures can coexist in some compounds.

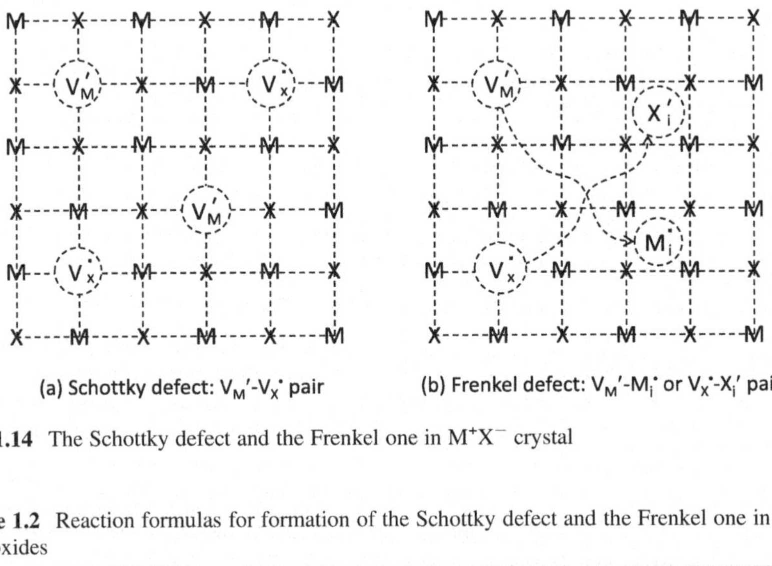

(a) Schottky defect: V_M'-$V_X^{\bullet}$ pair (b) Frenkel defect: V_M'-$M_i^{\bullet}$ or $V_X^{\bullet}$-X_i' pair

Fig. 1.14 The Schottky defect and the Frenkel one in M^+X^- crystal

Table 1.2 Reaction formulas for formation of the Schottky defect and the Frenkel one in halides and oxides

Compound	Defect equilibrium	Compound	Defect equilibrium
NaCl	$\text{null} = V'_{Na} + V^{\bullet}_{Cl}$	MgO	$\text{null} = V''_{Mg} + V^{\bullet\bullet}_{O}$
LiF	$\text{null} = V'_{Li} + V^{\bullet}_{F}$	CaO	$\text{null} = V''_{Ca} + V^{\bullet\bullet}_{O}$
CaF_2	$F^X_F = V^{\bullet}_F + F'_i$	UO_2	$O^X_O = V^{\bullet\bullet}_O + O''_i$
	$Ca^X_{Ca} = V''_{Ca} + Ca^{\bullet\bullet}_i$		$U^X_U = V''''_U + U^{\bullet\bullet\bullet\bullet}_i$
	$\text{null} = V''_{Ca} + 2V^{\bullet}_F$		$\text{null} = V''''_U + 2V^{\bullet\bullet}_O$
BeO	$\text{null} = V''_{Be} + V^{\bullet\bullet}_O$	AgBr, AgCl	$Ag^X_{Ag} = V'_{Ag} + Ag^{\bullet}_i$

1.3.3 Nonstoichiometric Defects

Compounds of typical elements have definite compositions depending on the ionic valence. On the contrary, transition metal compounds exhibit a variation of composition from the integer ratio of cation to anion, because the atomic valence of transition metal is variable. They are called "nonstoichiometric compound" and classified into the following four types. Here, "δ" means the deviation from stoichiometry.

Cation-deficiency type (cation vacancy): $Fe_{1-\delta}O$, $Co_{1-\delta}O$, $Ni_{1-\delta}O$, $Cu_{2-\delta}O$
Cation-excess type (interstitial cation): $Zn_{1+\delta}O$, $Cr_{2+\delta}O_3$, $Cd_{1+\delta}O$
Anion-deficiency type (anion vacancy): $ZrO_{2-\delta}$, $PrO_{2-\delta}$
Anion-excess type (interstitial anion): $UO_{2+\delta}$

Since the ionic radius of cation is smaller than that of anion, cation-deficiency and cation-excess types easily form. $Fe_{1-\delta}O$ has a large amount of cation vacancy as $\delta \leq 0.15$, but the defect amount is ordinarily small as $\delta < 0.01$ in $Co_{1-\delta}O$, $Ni_{1-\delta}O$, and $Zn_{1+\delta}O$. When the cation size is large, the anion-deficiency and anion-excess types are found because the lattice space expands. The oxygen vacancies form in $ZrO_{2-\delta}$. The formation of interstitial anion is rare as in $UO_{2+\delta}$.

The nonstoichiometry in transition metal oxides provides a large influence on the electronic properties. ZnO works as n-type semiconductor due to cation-excess defect. According to the Kröger-Vink notation, the equilibrium equation for the formation of interstitial cation is written as follows:

$$ZnO = Zn_i^{x} + 1/2O_2 \tag{1.1}$$

This equation means that oxygen is removed from stoichiometric ZnO, leaving Zn at interstitial site, where neutral Zn atom (Zn_i^{x}) is assumed although the formal cation is Zn^{2+} on lattice. The neutral Zn atom releases the first electron and then the second one as follows:

$$Zni^{x} = Zn_i^{\bullet} + e' \tag{1.2}$$

$$Zn_i^{\bullet} = Zn_i^{\bullet\bullet} + e' \tag{1.3}$$

There are three interstitial defects: Zn_i^{x}, $Zn_i^{\bullet}$ and $Zn_i^{\bullet\bullet}$, which depend on temperature and oxygen partial pressure (Po_2). The content of interstitial Zn increases at low oxygen pressure. The major interstitial defect is suggested by applying the mass action law to these equilibrium equations. When Zn_i^{x} is major, the content is proportional to $Po_2^{-1/2}$. When $Zn_i^{\bullet}$ is major, the content is proportional to $Po_2^{-1/4}$. When $Zn_i^{\bullet\bullet}$ is major, the content is proportional to $Po_2^{-1/6}$. According to Eqs. (1.2) and (1.3), the concentration of electron (n) is determined from the content of interstitial Zn: $n=[Zn_i^{\bullet}]$ or $n=2[Zn_i^{\bullet\bullet}]$. Therefore, the electrical conductivity depends on oxygen partial pressure. Figure 1.15 illustrates the schematic of the dependence

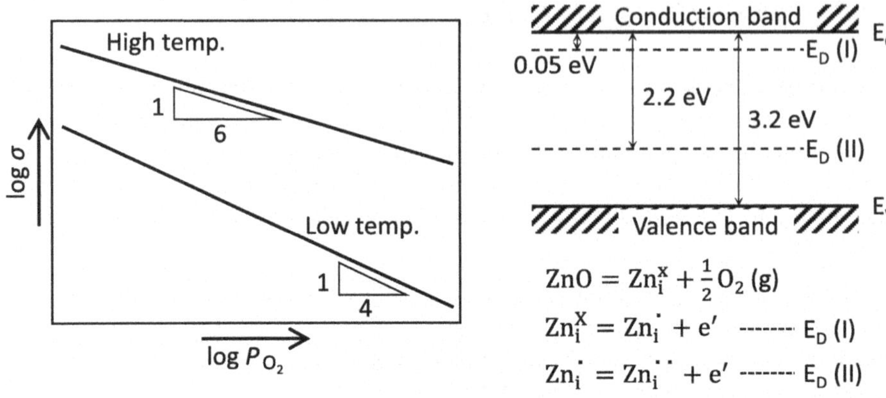

Fig. 1.15 Schematic of dependence of electrical conductivity on oxygen partial pressure and electron energy band model of ZnO

of electrical conductivity on oxygen partial pressure and electron energy band model of ZnO. The conductivity is proportional to $Po_2^{-1/4}$ at low temperature, indicating that the major defect is $Zn_i^{\bullet}$. At higher temperature, the proportionality to $Po_2^{-1/6}$ indicates the major defect of $Zn_i^{\bullet\bullet}$. Zn_i^{X} and $Zn_i^{\bullet}$ form shallow impurity level and deep one, respectively, within the bandgap between valence and conduction bands. The impurity levels work as electron donor, that is, electrons are excited from donor levels to conduction band.

CoO works as p-type semiconductor due to cation-deficiency defect, where vacancies form at Co sites. Some Co^{2+} ions change to Co^{3+} ions for charge compensation. The extra positive charges are not localized at specific Co sites and move as positive holes, resulting in the electrical conduction. Theoretically, Co vacancies are assumed to trap positive holes. It means that the vacancies are positively charged and release positive holes. According to the electron energy band model, the positively charged vacancies work as electron acceptor. Electrons in valence band are excited to the acceptor levels and leave positive holes as charge carrier in the valence band.

1.3.4 Defects in Solid Solution

Solid solution between definite compounds with different ionic valences forms point defects to keep charge balance. The solid solutions of c-ZrO_2 with CaO and Y_2O_3 are typical. The equilibrium equation in CaO-doped ZrO_2 is written as follows:

$$CaO = Ca_{Zr}'' + V_O^{\bullet\bullet} + O_O^{x} \tag{1.4}$$

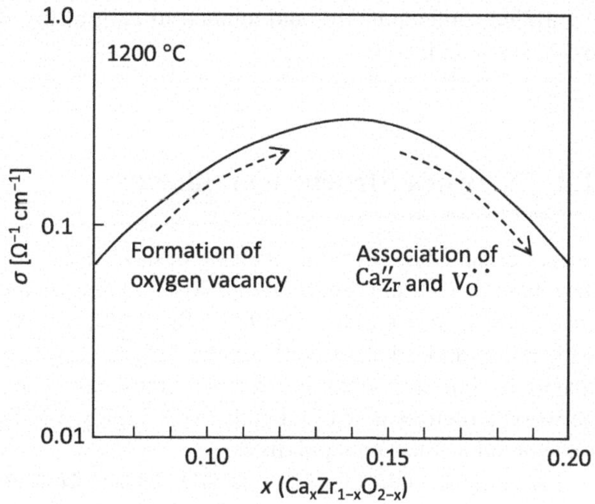

Fig. 1.16 Dependence of oxide ion conductivity on Ca content in $Ca_xZr_{1-X}O_{2-X}$

Ca^{2+} is substituted for Zr^{4+}. Since the positive charges are deficient, oxygen vacancies form. Through the oxygen vacancies, O^{2-} ions easily diffuse, resulting in oxide ion conduction. As shown in Fig. 1.16, the electrical conductance increases with an increase in Ca content owing to the formation of oxygen vacancies. However, it decreases at high Ca content. This is caused by the association between $Ca_{Zr}''$and $V_O^{\bullet\bullet}$ with the Coulombic force, because the associated vacancies cannot be used for oxide ion conduction.

Electrical conductivity of nonstoichiometric metal-oxide semiconductor can be controlled by doping of definite metal oxide with different ionic valences. The equilibrium equation in doping of Al_2O_3 to n-type semiconductor, ZnO, is written as follows:

$$Al_2O_3 = 2Al_{Zn}{}^{\bullet} + 2O_O{}^{x} + 2e' + 1/2O_2, \quad (1.5)$$

where Al^{3+} is substituted for Zn^{2+} with release of O_2. The concentration of electron increases by Al_2O_3 doping, leading to the increase in electrical conductivity of n-type semiconductor. The equilibrium equation in doping of Li_2O to p-type semiconductor, CoO, is written as follows:

$$Li_2O + 1/2O_2 = 2Li_{Co}{}' + 2h^{\bullet} + 2O_O{}^{x}, \quad (1.6)$$

where Li^{+1} is substituted for Co^{2+} with introduction of O_2. The Li_2O doping forms positive holes, which is the same meaning as disappearance of electrons. This causes the increase in electrical conductivity of p-type semiconductor. Higher-valence metal ions are doped to n-type semiconductor, while lower-valence metal ions are doped to p-type semiconductor. This type of semiconductor is called valence-controlled semiconductor, and applied to control the electrical properties of many oxide semiconductors.

1.4 Network Structure of Silicate

Ceramics have been developed on the basis of technologies of pottery and porcelain. The main components are SiO_2 and Al_2O_3. They are abundant in the earth's crust as seen in the Clarke number, O 49.5%, Si 25.8%, and Al 7.56%, and exist as silicates with alkali and alkaline earth oxides (Na_2O, CaO, etc.). Silicates are traditional materials, but important in ceramic industries. The feature is the polymerized network structure of Si-O bonds with highly covalent behavior, and they are understood as inorganic polymers.

Fundamental unit of silicate is SiO_4 tetrahedron, and they are connected at the corner of tetrahedron in various manners, forming network structures in one, two, and three dimensions. Typical structures are illustrated in Fig. 1.17. Connection mode of SiO_4 unit is ordered with O/Si atomic ratio ranging in $2 \leq O/Si \leq 4$. When $O/Si = 4$, SiO_4 exists as monomer. They are polymerized to chain, layer, and steric structure: one-dimensional chain or ring at $O/Si = 3$, two-dimensional layer at $O/Si = 2.5$, and pure SiO_2 at $O/Si = 2$. The polymerization degree is low at high O/Si ratio, including Na^+, K^+, Mg^{2+}, Al^{3+}, OH^-, etc. for charge compensation.

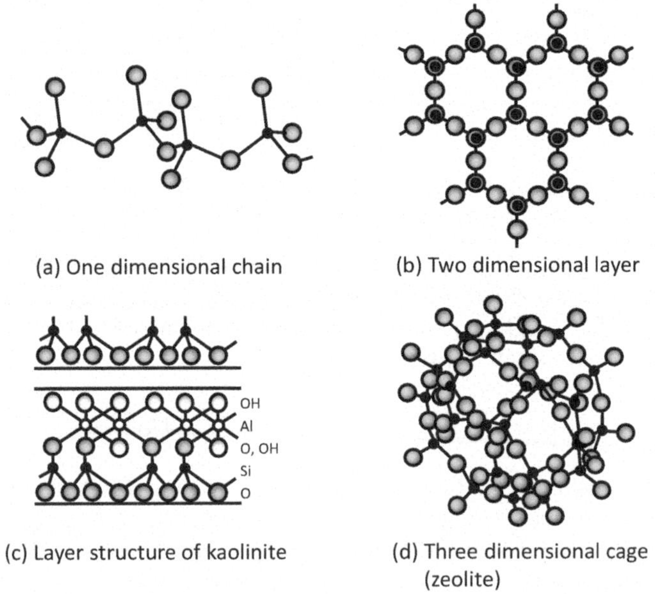

Fig. 1.17 Typical network structures of silicates by one-, two-, and three-dimensional arrangements

Chrysotile, $Mg_6(OH)_6Si_4O_{11}\cdot H_2O$, has double-chain structure. This is natural inorganic fiber and called asbestos. This has been widely used as heat-insulating material, but the use is now prohibited owing to the hazardous nature of sharp crystalline fiber. Kaolin has double-layer structure consisting of SiO_4 and AlO_6 layers. The composition is written as $(OH)_3Al_2(OH)\cdot Si_2O_5$ according to the layer structure. SiO_4 and AlO_6 layers are covalently bonded through oxygen. The van der Waals force works between the double layers. When kaolin is kneaded with water, H_2O molecules are intercalated between the double layers, causing plasticity. After drying, it solidifies due to removing H_2O molecules. This phenomenon has been applied for shaping of pottery and porcelain. The other layer structures are found in talc and mica. Mica is exfoliative. Mica films have been applied as insulating sheet in electrical parts. Zeolite forms spherical cage consisting of SiO_4 and AlO_4 tetrahedron units. This is featured for pore size at nanometer level in the network, providing various functions of molecular sieve, ion exchange, and adsorption of organic substances.

1.5 Glass Structure

Glass is amorphous solid which is frozen as it is from supercooled liquid. Typical inorganic glass is SiO_2-based one. The structural features of oxide glasses are summarized as follows:

(1) The number of cation bonded to oxygen is one or two: M-O and M-O-M bonds are allowed.
(2) The oxygen coordination number of cation is 4 or less.
(3) Oxygen polyhedron shares corner, but does not share edge and face.
(4) Oxygen polyhedron should share at least 3 corners.

The conditions, (1), (2), and (3), are required to allow flexibility of glass structure, but the condition, (4), is required to hold the structure. The covalent valence number of oxygen is two, but the number of cation bonded to oxygen can become three or over in ionic crystals, causing rigid framework. The bond angle at the corner of SiO_4 tetrahedron is flexible, allowing amorphous random structure of glass.

Glass components are classified into glass former, intermediate and modifier. Glass former is fundamental component to constitute glass: SiO_2, B_2O_3, P_2O_5, etc. Intermediate works as a part of glass former and improves the property of glass: Al_2O_3 and ZnO improve chemical stability, and PbO increases refractive index to produce crystal glass. Modifier lowers melting point and viscosity: Na_2O, K_2O, CaO, etc., because Si-O networks are cut off as following equation:

$$-\overset{|}{\underset{|}{\mathrm{Si}}}-\mathrm{O}-\overset{|}{\underset{|}{\mathrm{Si}}}- + \mathrm{Na_2O} \rightarrow 2\left[-\overset{|}{\underset{|}{\mathrm{Si}}}-\mathrm{O}^-\right] + 2\mathrm{Na}^+ \tag{1.7}$$

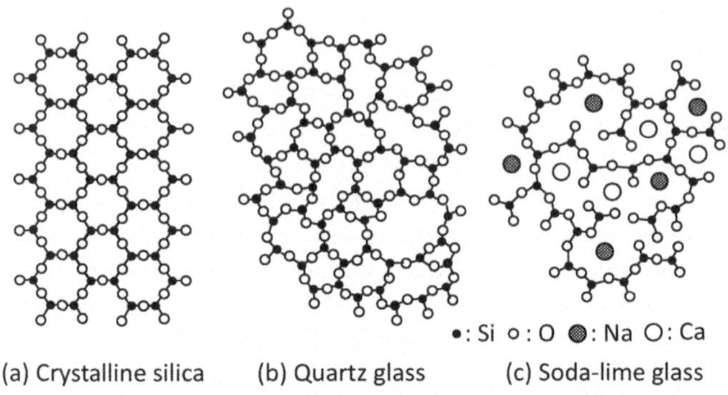

Fig. 1.18 Network structures of crystalline silica and amorphous silica glasses (2D model)

Network structures of glasses are illustrated in Fig. 1.18. Silica crystal has regular structure. Quartz glass is pure glass. It is produced by fusion of quartz crystal, and called fused quartz or fused silica. SiO_4 units are randomly connected, which is different from regular arrangement in quartz. The quartz glass has been applied as high-grade refractory glass because of high melting point, low thermal expansion coefficient, and high thermal shock resistance. However, it is expensive. Soda-lime glass includes Na_2O and CaO as modifier. The composition of practical soda-lime glass is SiO_2 65–75, Al_2O_3 0.5–4, CaO 5–15, and Na_2O 10–20 in wt%. Na^+ and Ca^{2+} are incorporated into SiO_2 network. The low melting point makes it easy to produce usual flat glasses. Borosilicate glass includes B_2O_3. The addition of B_2O_3 to SiO_2 network weakens the glass because the number of bonds decreases by doping of BO_3 triangle compared to SiO_4 tetrahedron. The simultaneous addition of Na_2O and B_2O_3 forms BO_4 tetrahedron, yielding stable glass. The composition of practical borosilicate glass is SiO_2 65–75, B_2O_3 5–12, Al_2O_3 1–5, CaO 5–8, Na_2O 8–14, and K_2O 1–5 in wt%. Since the borosilicate glass is superior in thermal and chemical stability, it has been used as laboratory glasses and kitchen utensils. The famous commercial name is "Pyrex."

1.6 Highly Ordered Structure by Dimensional Control

Materials are constructed from atoms or molecules by chemical bonding to form crystal or amorphous structure. Crystals and amorphous phases form assembled structures and bring out their functions. From the practical viewpoint, morphological control is important for their applications. The morphology of material is hierarchically classified as follows: zero-dimension (particle), one-dimension (fiber), two-dimension (film), and three-dimension (bulk). They are totally termed the highly ordered structure. Table 1.3 summarizes highly ordered structures: morphology, characterization, and application.

Table 1.3 Highly ordered structures: morphology, property and application

Dimension	Morphology	Material	Property	Application
Zero-dimensional control	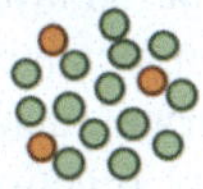	Powder Particle	Size effect Surface effect	Conducting fluid Magnetic particle Phosphor, catalyst
One-dimensional control		Fiber Whisker	High thermal stability High strength Transparency	Composite filler Optical fiber
Two-dimensional control		Thin film Thick film	Surface modification Electrical property Optical property	Heat insulating coating Electronic device Solar cell
Three-dimensional control	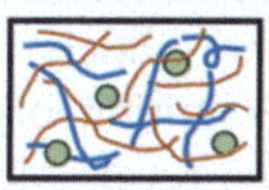	Sintered body Composite Organic/inorganic hybrid	High strength Heat resistance Wear resistance Electromagnetic property	Structural part Engine part Machining part Electronic device
Spatial control	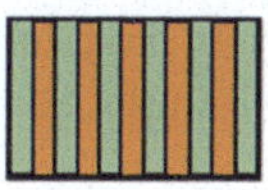	Porous material Layered compound Zeolite	Molecular sieve Molecular recognition Separation/ adsorption	Filter, Gas sensor Catalyst

Fine ceramic particles have been required to produce high-density sintered bodies because of its high sinterability. Dispersive particles are desired as filler of polymer composite to improve mechanical strength and thermal conductivity. High specific surface area enhances catalytic activity. Nano-sized particles are expected to exhibit quantum size effect in electromagnetic and optical properties. Ceramic fibers have been applied for reinforcement of composite with polymer, metal, and ceramics. Carbon and SiC fibers with high strength and heat resistance are typical. There are long fiber, short fiber, and whisker. SiC whisker has high strength. SiC-based long fibers with high heat resistance have been developed. Quartz glass fiber has been widely used for optical telecommunication. Carbon nanotube and graphene attract great attention as new materials in the fields of battery, capacitor, photocatalyst, and so on. Ceramic thin films and thick ones play an important role for surface modification to protect metal substrate. Thermal-barrier coating on refractory alloy is promising in application of high temperature engineering materials. Functionally gradient materials with graded composition are effective to relax the stress between substrate and coating. The functions of coated films themselves are unique due to their anisotropic size effect in electronic devices.

Fine-grained microstructure is required to enhance the mechanical strength of sintered ceramics. Ceramic composites have widened their application by combining different properties. Si_3N_4 is high strength material. The addition of soft h-BN is effective to provide machinability and improve thermal shock resistance. The addition of electrically conductive TiN enables electrical discharge machining. Microstructural control at nanometer level creates novel functions. This is named nanocomposite. Si_3N_4-BN nanocomposite exhibits extremely high thermal-shock resistance compared to the conventional composite. Organic-inorganic hybrid is attractive due to combination of flexibility of organics and rigidity of inorganic substance. Sol-gel method is useful to polymerize organic and inorganic frameworks. Hybrid of siloxane system is famous for hard contact lens. Layer compound like clay is applicable to produce polymer composites. The layers are exfoliated and dispersed by incorporation of polymer such as nylon and polystyrene to produce polymer nanocomposites.

Porous ceramics are superior in durability and applied for water purification and exhausted gas treatment. Sintered porous ceramics are widely used. The size and amount of pores can be controlled by sintering condition of powder compact. Mesoporous materials have small pore size at meso-level. Mesoporous silica is typically synthesized by sol-gel method. Layer compound and zeolite are microporous materials at molecular level. Fine porous materials have unique functions like molecular sieve and molecular recognition and can be applied for gas separation, sensor, and catalyst.

Further Readings

1. W. D. Kingery, H. K. Bowen, D. R. Uhlmann, *Introduction to Ceramics*, 2nd edn. (Wiley, New York, 1976); W. Komatsu, T. Sata, Y. Moriyoshi, K. Kitazawa, K. Uematsu, *Introduction of Materials Science in Ceramics: Basic Edition*, Japanese Translation (Uchida Rokakuho Publishing Co., Ltd., Tokyo, 1980)
2. F. S. Galasso, *International Series of Monographs in Solid State Physics, Vol. 7: Structure and Properties of Inorganic Solids* (Pergamon Press Ltd., Oxford, 1970); M. Kato, K. Uematsu, *Crystal Chemistry of Fine Ceramics: Structure and Property of Inorganic Solid Compounds*, 3rd Japanese Translation (AGNE Gijutsu Center Inc., Tokyo, 2002)
3. Y. Arai, *Materials Chemistry of Ceramics* (Dainippon Tosho Publishing Co., Ltd., Tokyo, 1980)
4. H. Yanagida, K. Koumoto, M. Miyayama, *Chemistry of Ceramics*, 2nd edn. (Maruzen Publishing Co., Ltd., Tokyo, 1993)
5. Y. Moriyoshi, T. Sasamoto, K. Uematsu, Y. Ikuma, *Basic Science of Ceramics*, 4th edn. (Uchida Rokakuho Publishing Co., Ltd., Tokyo, 1995)
6. J. Shiokawa, *Introduction to Inorganic Materials* (Maruzen Publishing Co., Ltd., Tokyo, 1996)
7. T. Endo, H. Iwasaki, T. Tsurumi, A. Nakahira, H. Takizawa, *Introduction to Crystal Chemistry* (Kodansha Scientific Ltd., Tokyo, 2000)

Chapter 2
Phase Equilibrium and Phase Diagram

Junichi Tatami

Abstract Phase equilibrium and phase diagram advise us what is the stable substance under certain circumstances and what condition is needed for the stability of the substance. This chapter describes the fundamental to understand the phase equilibrium and the phase diagram: the phase rule, the phase transformation with the change in the temperature or the pressure, the relationship between the temperature and the pressure in the phase transformation (so-called the Clausius–Clapeyron equation), and the phase change in compositional variation. The ways of reading and understanding of one-, two-, and three-component phase diagrams are explained using several practical phase diagrams.

Keywords Phase equilibrium · Phase diagram · Phase rule · Phase transformation · Eutectic reaction · Peritectic reaction

2.1 Fundamental of Phase Diagrams

A phase diagram is a map that indicates which phases are present when substances are in equilibrium with the surrounding environment [1–5]. An equilibrium state is achieved for the combination of phases that have the minimum total free energy, that is, the most stable system in the environment of interest. Therefore, a phase diagram is an illustration of the trajectory of the stable phases, and it gives the number and the composition of the phases present, as well as their relative proportions, as a function of temperature, pressure, and composition.

A phase is any portion distinguished by an interface and has homogeneously and specifically physical and chemical properties. It is classified by the basic state of matter: solid, liquid, or gas. For example, in a glass of iced water, the ice and water are different phases. A system composed of two or more phases is described as a

J. Tatami (✉)
Graduate School of Environment and Information Sciences, Yokohama National University, Yokohama, Japan
e-mail: tatami-junichi-xv@ynu.ac.jp

J. Hojo (ed.), *Materials Chemistry of Ceramics*,
https://doi.org/10.1007/978-981-13-9935-0_2

heterogeneous system, whereas a system composed of only one phase is termed a homogeneous system. A system composed of gases is always a homogeneous system because the gases are completely mixable in each other. A liquid can be composed of only one phase, such as a mixture of water and ethanol, or two or more phases, such as a system of water and oil. Solids are treated in the same way as liquids. In particular, crystalline solids having different crystal structure are regarded as different phases, even if their chemical composition is the same. For example, diamond and graphite are different allotropes of carbon. This is called polymorphism. A solid in which different atoms or ions are dissolved is a single phase, as is the case with a liquid phase, and is called a solid solution. For example, NiO and MgO form a solid solution because both have cubic crystal structures, the valences of Ni and Mg are equal, and the difference in the ionic radius of Ni^{2+} and Mg^{2+} are within $\pm 15\%$. In this case, the solute atoms or ions are substituted by solvent atoms or ions. This is called a substitutional solid solution. Alternatively, smaller atoms or ions can dissolve into the interstices of the host crystal, thus forming an interstitial solid solution.

2.2 Phase Rule and Phase Transformation

2.2.1 Gibbs Phase Rule

In a heterogeneous system at equilibrium, the relationship between the number of phases and the components of the system is determined by the following phase rule, as proposed by Josiah Willard Gibbs [6]:

$$F = C - P + 2 \tag{2.1}$$

Here, F is the number of degrees of freedom, C is the number of components, and P is the number of phases. This rule remains true as long as the system is at equilibrium, even in complex cases. For example, when ice, water, and water vapor coexist, the chemical composition of each phase is H_2O; thus, it is a one-component system, not a two-component system of H and O. If H_2 gas is added to this system, it becomes a two-component system of H_2O and H_2. Furthermore, when a system is composed of $CaCO_3$ and its decomposition products (CaO and CO_2), it is also two-component system because choosing any two of the three phases fixes the composition of the remaining phase, even if all three phases are present at equilibrium. Depending on the number of components, a one-component phase diagram or two-component phase diagram is drawn. The number of degrees of freedom is the number of independent variables, such as pressure, temperature, and concentration of the components, without changing the number of phases. In general, a phase diagram is illustrated on the projected temperature–concentration or pressure–temperature coordinate plane.

The Gibbs phase rule is derived as follows. To determine the concentration of each phase, $C-1$ concentration terms are required for each phase because the sum of concentration terms is 100%. In the whole system, the number of concentration terms is $P(C-1)$ in total, considering the number of phases. Because the temperature and pressure must be specified, the total number of variables is $P(C-1) + 2$. There are $P-1$ equilibrium equations for each component; thus, the total number of equations is $C(P-1)$. Consequently, the number of independent variables (i.e., the number of degrees of freedom, F) becomes $P(C-1) + 2-C(P-1) = C - P + 2$.

2.2.2 Phase Transformation

Melting, solidification, sublimation, vaporization, and polymorphic transitions result from changes in the temperature or pressure. Such state changes are termed phase transformation. Because the state having minimum free energy at a certain temperature and pressure condition is the stable phase, a phase diagram can be drawn by estimating the stable region of each phase from the change in the free energy. Here, the phase transformation between phase α and β on temperature change is considered. The Gibbs free energy is given as follows:

$$G = H - TS = U - TS + PV \quad (2.2)$$

where G is the Gibbs free energy, H is the enthalpy of the system, T is the absolute temperature, S is the entropy of the system, U is the internal energy, P is the pressure on the system, and V is the volume of the system. The enthalpy, H, dominates the Gibbs free energy at lower temperatures, and TS plays an important role at higher temperatures. When the free energies of phases α and β and G_α and G_β intersect at temperature T_0, as shown in Fig. 2.1a, both phases have the same stability because the free energy is the same at T_0. If phase β has a larger absolute value of the slope of the free energy ($(\partial G/\partial T)_P = -S$)) under constant pressure, P, at temperature, T_0,

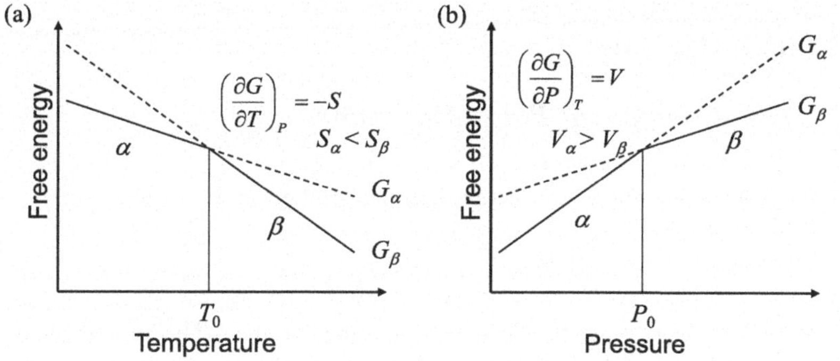

Fig. 2.1 Temperature and pressure dependence of free energy and phase transformation

than phase α, the phase α is stable at temperatures lower than T_0, and phase β is stable at higher temperatures; that is, a phase transformation from α to β occurs at temperature T_0. At this point, latent heat ($L = T_0(S_\beta - S_\alpha)$; S_α and S_β are the entropy of phase α and β, respectively) must be provided for the phase transformation. Phase transformation by pressure change can be considered in the same way as for temperature change (Fig. 2.1b). Because the free energies of phases α and β are equal at pressure P_0 and constant temperature T, a phase transformation from α to β occurs. The slope at pressure P_0 ($(\partial G/\partial P)_T = V$) is equal to the volume (V_α and V_β are the volume of phase α and β, respectively) of the system, which means that the phase having the smaller volume (i.e., higher density) is stable at higher pressure when the volume changes with the phase transformation.

The enthalpy and density change at the phase transformation point, and this is called a first-order phase transformation because the first derivative of the free energy is discontinuous. The melting of solids, vaporization of liquids, and polymorph change are first-order phase transformations. On the other hand, the phase transformation in which the second derivative of the free energy is not continuous is called a second-order phase transformation, such as magnetic phase transformation and superconducting transformation.

In the case of a first-order phase transformation, the relationship between the temperature and the pressure for the phase transformation is described by the Clausius–Clapeyron equation:

$$\frac{dP}{dT} = \frac{\Delta H}{T\Delta V} = \frac{\Delta S}{\Delta V} \tag{2.3}$$

where ΔH, ΔS, and ΔV are the changes in the enthalpy, entropy, and volume accompanying the phase transformation. This relationship indicates the slope of the boundary line between two phases and, like the Gibbs phase rule, is an important equation. Ordinarily, the melting of solid and the vaporization of liquid occur with the endothermic change ($\Delta H > 0$) and the volume increase ($\Delta V > 0$), resulting in positive slope of pressure vs. temperature line in one-component phase diagram.

2.3 Phase Diagram

2.3.1 One-Component Phase Diagrams

The Gibbs phase rule for a one-component phase diagram ($C = 1$) is

$$F = 1 - P + 2 = 3 - P \tag{2.4}$$

The number of degrees of freedom is zero, one, or two. The two valuables are assigned to pressure and temperature. The composition is not important because only one component is considered. Figure 2.2 shows a typical one-component

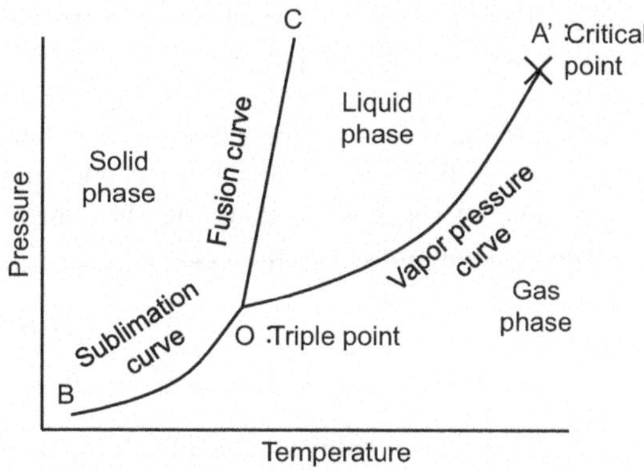

Fig. 2.2 Typical one-component phase diagram

phase diagram. In the region where only one phase exists, the number of degrees of freedom is two ($F = 3 - 1 = 2$), and the number of phases does not change, even if the pressure and temperature change slightly; that is, the pressure and temperature can be varied independently. On the boundary line between two phases, both phases coexist (for instance, there are solid and gas phases on line BO); that is, the number of phases is two. In this case, the number of degrees of freedom is one ($F = 3 - 2 = 1$), and, when the temperature or the pressure is changed, the other valuable is determined dependently. The boundary lines between solid and liquid, liquid and vapor, and solid and vapor are called the fusion curve (or the melting point curve, line CO), the vapor pressure curve (line A'O), and the sublimation curve (line BO), respectively. The point where three phases coexist in a one-component system ($P = 3$) under equilibrium conditions, as is the case for point O in Fig. 2.2, is called the triple point. The temperature and the pressure are invariant, and the point is also called an invariant point because the number of degrees of freedom is zero ($F = 3 - 3 = 0$). Furthermore, there is no longer a difference between the liquid and gas phases at a higher temperature and pressures beyond point A$'$, which is called the critical point.

A phase diagram can be determined from some physical relationships between the phases. A hypothetical construction contributes strongly to better familiarity with phase diagrams and their understanding. Thus, the following exercise is considered:

- A one-component system consisting of two solid phases (S_1 and S_2), one liquid phase (L), and one vapor phase (V).
- Density: $S_1 > S_2 > L$.
- S_1 and S_2 are melted and sublimated.
- The phase transformation from S_1 to S_2 is endothermic.

This phase diagram has boundary lines between the liquid and solid phases and between the vapor and solid phases; that is, there are two melting point curves and two sublimation curves. Furthermore, one vapor pressure curve and a boundary line between S_1 and S_2 also exist.

The slopes of the lines can be determined from the Clausius–Clapeyron equation (Eq. 2.3). In the phase transformation from S_1 to S_2, the reaction is endothermic

($\Delta H > 0$) and the volume expands ($\Delta V > 0$). This means the slope of the boundary line between S_1 and S_2 is positive. The slope of the other boundary line can also be estimated.

Because the volume of the vapor phase is much larger than those of the solid and liquid phases, the volume change for vaporization or sublimation is nearly equal to the volume of the gas. In addition, the vapor behaves as an ideal gas, and the Clausius–Clapeyron equation becomes

$$\frac{dP}{dT} = \frac{\Delta H_V P}{RT^2} \tag{2.5}$$

where ΔH_V is the enthalpy change of sublimation or vaporization. By integrating this equation, the sublimation and vapor pressure curves are derived as

$$P = \exp\left\{-\left(\frac{\Delta H_V}{RT}\right)\right\} \tag{2.6}$$

This means that the sublimation pressure and the vapor pressure are expressed by the exponential function, resulting in a concave pressure vs. temperature line. On the other hand, the fusion curve and the boundary line of phase transformation between solid phases are straight because the volume change is very small. At the triple point, the slope of the fusion curve is larger than the sublimation curve, considering the volume change ($\Delta V_{fusion} << \Delta V_{sub}$), and the slope of the sublimation curve is larger than the vapor pressure curve, considering the enthalpy change ($\Delta H_{sub} > \Delta H_{vapor}$), in the Clausius–Clapeyron equation. Consequently, the phase diagram is as that shown in Fig. 2.3.

Figure 2.4 shows the phase diagram of water. The melting and boiling points of water at 1 atm are 0 and 99.974 °C, and the triple point is at 0.01 °C and 0.006 atm.

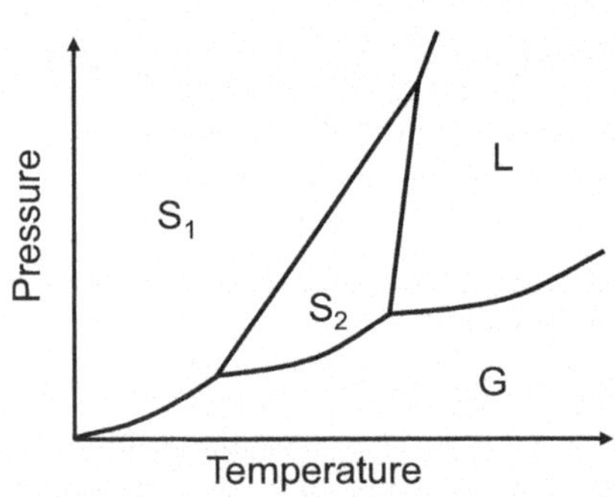

ΔH	S_1	S_2	L
S_1		+	+
S_2	−		+
L	−	−	

ΔV	S_1	S_2	L
S_1		+	+
S_2	−		+
L	−	−	

Fig. 2.3 Phase diagram determined from the given condition. Tables indicate the change in the enthalpy and volume accompanying the phase transformation. The ΔH and ΔV of solid and liquid to gas phase are always positive

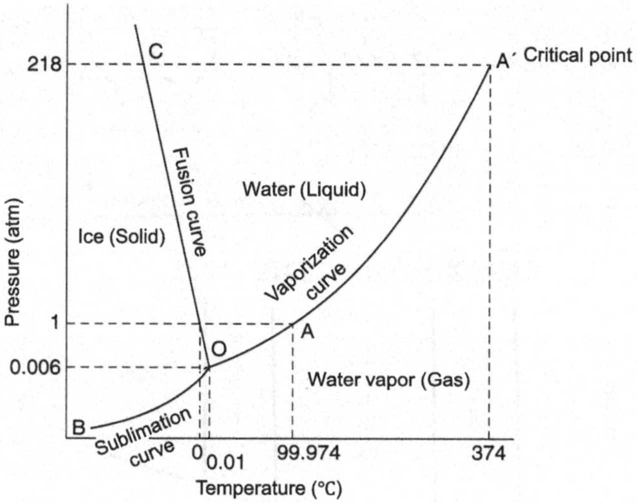

Fig. 2.4 Phase diagram of water

This phase diagram is introduced as a typical one-component phase diagram. However, the slope of the fusion curve is exceptionally negative because the density of liquid water is larger than that of ice near its melting point, as can be seen from the floating of ice on water. In this case, the ice melts on the application of pressure. This is the reason for being able to skate on ice: water forms under the ice skate blade because of the pressure, and the liquid turns back to ice after the release of the pressure. The triple point is important to decide the physical state of matter. Carbon dioxide is frozen directly from gas without going through liquid phase, because the pressure at the triple point is above 1 atm. This is the dry ice.

Figure 2.5 shows the phase diagram of SiO_2. As shown, there are many polymorphs. The solid lines indicate boundaries between equilibrium phases, and the dashed lines show metastable phases which exist during cooling. Molten SiO_2 is frozen in glassy state. High-temperature phases sometimes appear instead of the equilibrium phase, which is explained by the difference in the crystal structure between polymorphs. The phase transformation from α- to β-quartz occurs rapidly because of similarity of their structures. On the other hand, because the crystal structure of tridymite differs from that of cristobalite and a major change in the structure accompanied by significant atomic migration is required, rapid phase transformation from cristobalite to tridymite is difficult. The transformation from tridymite to quartz is also difficult. In these cases, the high-temperature phases of cristobalite and tridymite remain during cooling and then transform to the low-temperature phases.

Figure 2.6 shows the phase diagram of ZrO_2. Although the slopes of the boundary lines of the liquid/cubic phases and cubic/tetragonal phases are positive, that of the tetragonal/monoclinic phases is negative. This means that the volume expansion

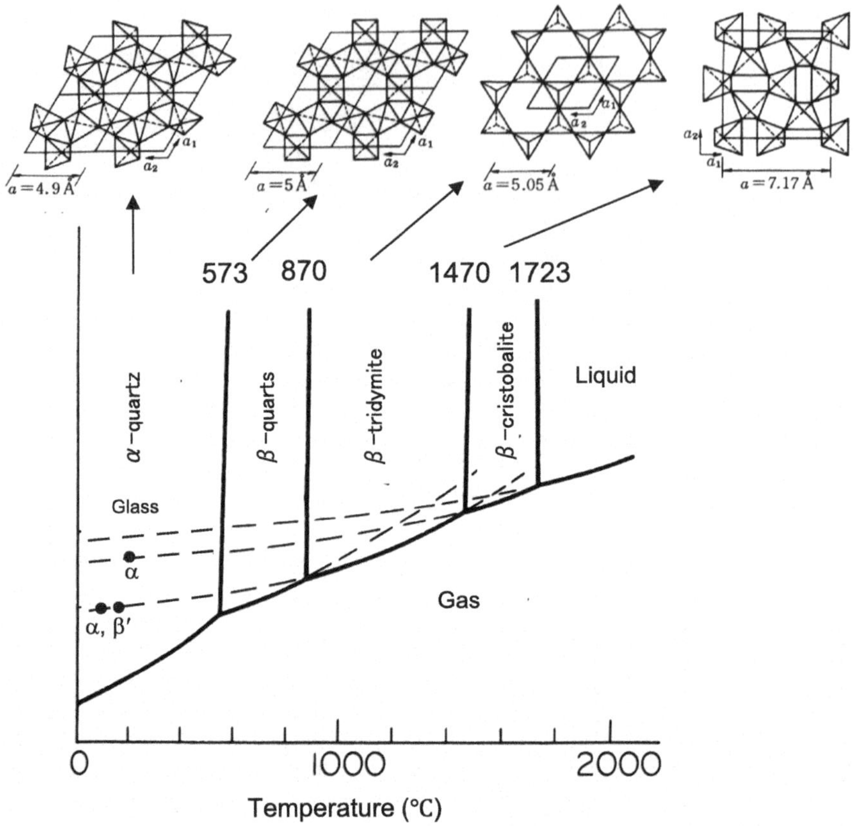

Fig. 2.5 Phase diagram and crystal structure of SiO_2. β-Cristobalite and β-tridymite are transformed to α and β' as metastable phases at low temperatures

occurs in the phase transformation from the tetragonal phase to monoclinic phase. In the case of pure ZrO_2, the sintered body is fractured in the cooling process because of the large volume change. However, tetragonal ZrO_2 ceramics can be obtained by adding CaO or Y_2O_3 because the additives make solid solutions and stabilize the tetragonal phase, even at room temperature. This material, known as tetragonal zirconia polycrystal (TZP) or partially stabilized zirconia (PSZ), shows high fracture toughness and strength because compressive forces arising from volume expansion are applied to the crack front.

Figure 2.7 shows the phase diagram of carbon. The thermodynamically stable phase at room temperature and ambient pressure is graphite. Diamond is the stable phase at higher temperatures and higher pressures and is naturally formed under high pressures and high temperatures in deep underground. This means that diamond is the metastable phase at room temperature and ambient pressure. Because there is large difference between the crystal structures, diamond almost never converts to graphite in the ordinary circumstance.

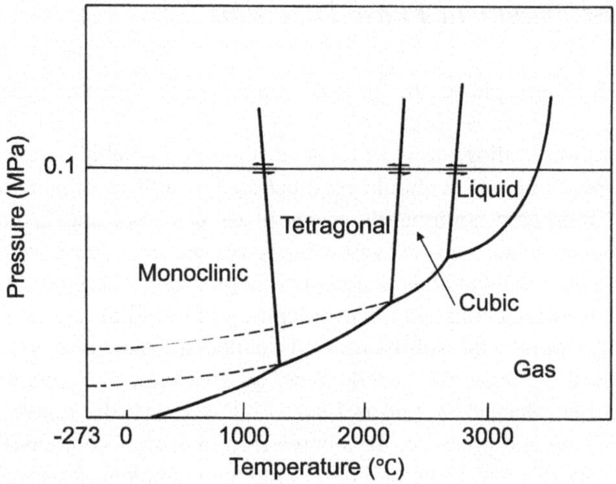

Fig. 2.6 Phase diagram of ZrO_2

Fig. 2.7 Phase diagram of carbon

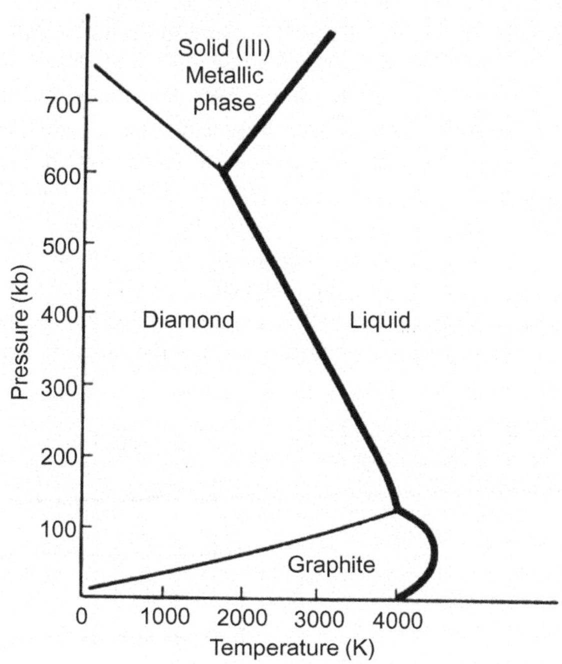

2.3.2 Two-Component Phase Diagrams

According to the Gibbs phase rule for two-component system, the maximum number of degrees of freedom is three ($F = 2 - P + 2 = 4 - P$). Therefore, the two-component phase diagram is influenced by three valuables, temperature, pressure, and composition, and it should be illustrated as a three-dimensional figure. In most cases of inorganic compounds, a gas phase does not appear until very high temperatures are reached and the vapor pressures are low. Furthermore, inorganic compounds are processed at almost atmospheric pressure. Therefore, it is convenient to treat the pressure as a constant in consideration of only liquid and solid phases. A system of only liquid and solid phases is called a condensed system. Because pressure is used as one of the degrees of freedom, the phase rule of the two-component condensed system is $F = C - P + 1$, and the number of variables is two: temperature and pressure. This is referred to as the "condensed phase rule."

Figure 2.8 shows an example of a two-component condensed phase diagram. The temperature and composition are placed on the vertical and horizontal axes, respectively. The liquidus line is the boundary line above which only a liquid phase exists, and no crystal can exist in this region. This means that the primary crystal begins to appear on cooling the melt under equilibrium condition at the liquidus line. On the other hand, below the solidus line, no liquid phase can exist. In the phase diagram, an isothermal line can be drawn between the compositions of two phases that are in equilibrium at that temperature. This line is called a tie line. Therefore, the composition of the liquid phase can be estimated from the liquidus line by drawing a tie line. In Fig. 2.8, solid phases are pure A and B, and the mixture of A and B appears under the solidus line. As shown in the Figs. 2.9 and 2.10, when the solid phase varies in composition at the equilibrium, the composition of the solid phase is also expressed by the intersecting point between the solidus line and the tie line. When only a single phase exists, the number of degrees of freedom is two ($F = 2 - 1 + 1 = 2$), and the temperature and the composition are the independent valuables. For example, in the

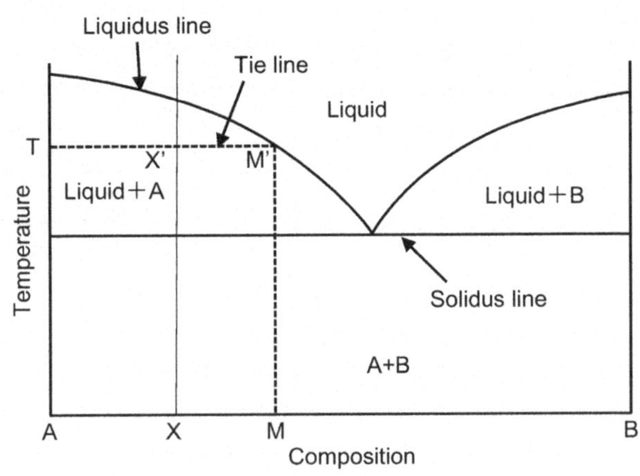

Fig. 2.8 An example of a two-component condensed phase diagram

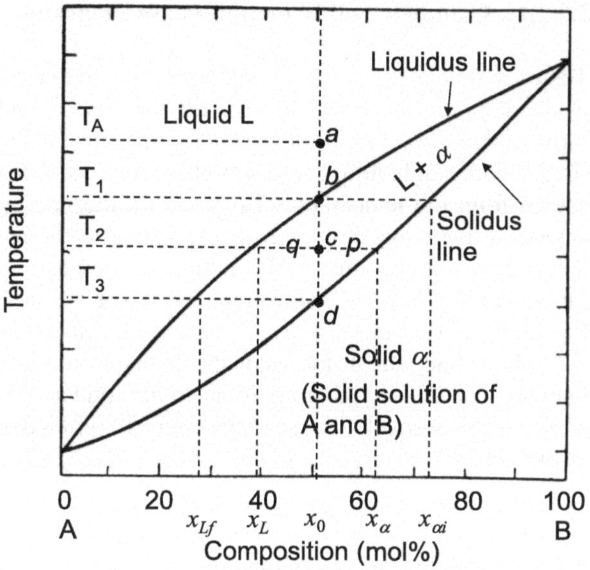

Fig. 2.9 Complete solid-solution phase diagram of A and B system

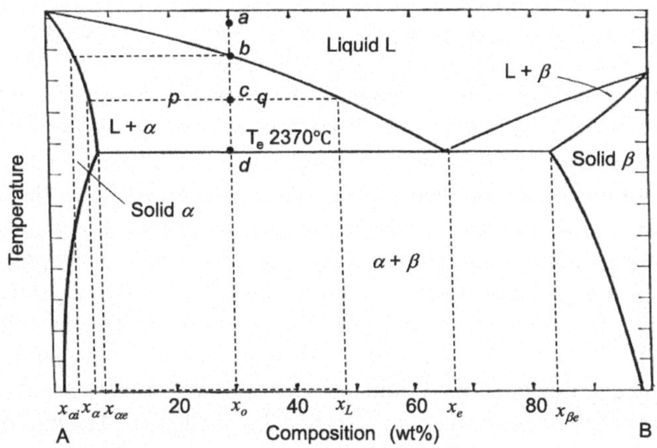

Fig. 2.10 Eutectic phase diagram of A and B system

region of the liquid phase alone, the number of phases is one, even if the composition and temperature change slightly. If both liquid and solid phases exist, such as A + Liquid or Liquid + B, the number of independent valuables is one ($F = 2 - 2 + 1 = 1$). This means that the solid and liquid compositions are automatically fixed when the temperature is set. The relative proportion of each phase is derived from the mass balance. For example, the proportion of the liquid phase to the solid phase at the average composition X in Fig. 2.8 can be expressed as AX/XM. This is called the lever principle from mechanical analogy to the mass balance calculation in the lever.

2.3.2.1 Complete Solid-Solution Phase Diagrams

There are several types of two-component phase diagram. A complete solid-solution phase diagram, as shown in Fig. 2.9 between A and B, is characterized by the formation of a solid solution at all compositions. This phase diagram is composed of single solid, single liquid, and two-phase regions. When a melt of composition x_0 is cooled from the temperature T_A (point a), the primary crystal of composition $x_{\alpha i}$ appears at the temperature T_1 corresponding to the intersection (point b) of the tie line and the liquidus line. At this point, the composition of the liquid phase is x_0. By further cooling to point c at the temperature of T_2, the compositions of the solid and liquid phases become x_α and x_L, which are estimated from the tie line, liquidus line, and solidus line. The distances on the tie line from point c to the liquidus and solidus lines are p and q, and the proportions of the solid and liquid phases are $q/(p + q)$ and $p/(p + q)$, respectively. At the temperature T_3 (point d), there is no liquid phase, only a solid phase.

2.3.2.2 Eutectic Phase Diagrams

Figure 2.10 shows the eutectic phase diagram of A and B. Phases α and β are solid solutions of A and B into which B and A are dissolved, respectively. When a melt of composition x_0 is cooled from the point a, the primary crystal of composition $x_{\alpha i}$ forms at point b on the liquidus line. With further cooling, phase α continuously precipitates. The compositions of the liquid and solid phases change along the liquidus and solidus lines, and their proportions are estimated by the lever principle, as is the case with the complete solid-solution phase diagram. For example, at point c, the liquid and solid phases of composition x_L and x_α coexist in the proportion of $p/(p + q)$ and $q/(p + q)$, respectively. At temperature T_e, the two crystals of composition $x_{\alpha e}$ and $x_{\beta e}$ precipitate from the melt of composition x_e. The eutectic reaction is expressed as $L \rightleftarrows \alpha + \beta$. This is a eutectic reaction, and it is an invariant reaction. During the eutectic reaction, one liquid and two solid phases coexist at equilibrium, and the number of degrees of freedom is 0 ($F = 2 - 3 + 1 = 0$). The microstructure of the solid phase formed at the eutectic point is characteristic. T_e and x_e are called the eutectic temperature and the eutectic composition, respectively. The most commonly observed microstructure is a fine lamellar structure of alternating layers of phase α and β. The cooling rate decreases because of the exothermic heat evolution of solidification, and the temperature during the eutectic reaction is constant until the reaction is completed. This phenomenon is utilized to prepare phase diagrams experimentally.

2.3.2.3 Congruent Melting-Type Phase Diagram

Intermediate compounds of two or more components frequently form in a system. Phase diagrams including intermediate compounds are classified concerning their melting behavior. One type of phase diagram including intermediate compounds is the congruent melting-type phase diagram, as shown in Fig. 2.11. In this phase diagram, intermediate compound AB melts directly to a liquid having the same composition as the solid. The congruent melting-type phase diagram may be regarded as a combination of two simple eutectic phase diagrams of A-AB and AB-B systems. $T_{A\text{-}AB,e}$ and $T_{AB\text{-}B,e}$ are eutectic temperatures of A-AB and AB-B, respectively.

2.3.2.4 Peritectic Phase Diagram

Figure 2.12 shows a typical peritectic phase diagram, which appears when the composition of the liquid phase in the three-phase equilibrium is not between those of the solid phases at equilibrium. Phases α and β are solid solutions of A and B. T_p is the peritectic temperature and $x_{\beta p}$ is peritectic composition. At temperature T_p, phase β forms in addition to phase α, and the liquid phase L of composition x_{Lp}, the solid phase α of composition $x_{\alpha p}$, and the solid phase β of composition $x_{\beta p}$ coexist. Since, at this point, the quantity of component B is in the sequence of α < β < L, the liquid phase is lack of component A to precipitate phase β. Eventually, the phase β is formed by the reaction between liquid phase L and solid phase α to compensate the shortage of component A in the liquid phase L. As a result, phase β is

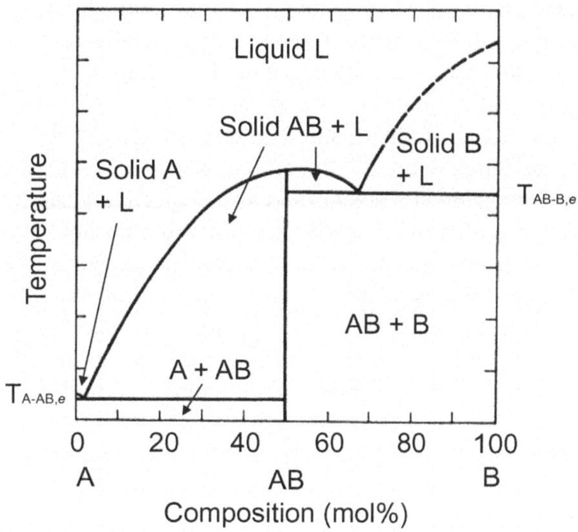

Fig. 2.11 Congruent melting-type phase diagram of A and B system

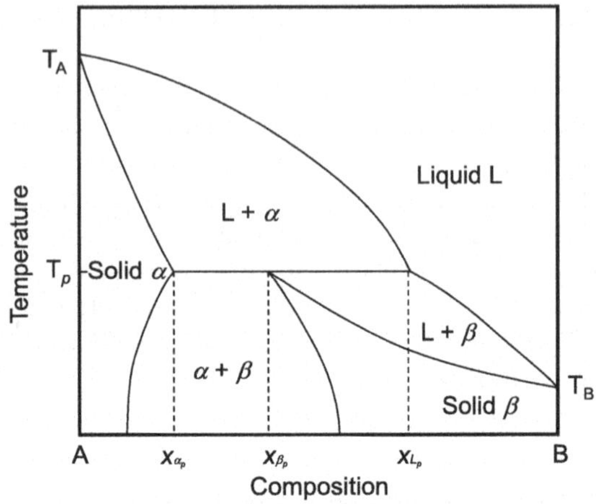

Fig. 2.12 An example of peritectic phase diagram

precipitated at the interface between the melt and phase α in a way to cover phase α and form peritectic crystals. The peritectic reaction is expressed as L + α ⇄ β.

When there is an intermediate compound at a composition below x_{Lp} instead of solid solution β, the phase diagram is also peritectic. The solid phase β decomposes to form liquid and solid phases having different compositions from the original intermediate compound.

2.3.2.5 Phase Diagram Including Immiscible Liquids

Liquids are not always soluble to one another, for example, water and oil. Phase diagram for system containing immiscible liquids is also possible, as shown in Fig. 2.13. This is the monotectic phase diagram. The area ACB in the figure is two-liquid region. When a liquid of composition X is cooled, two liquid phases appear at the point C. The compositions of the two liquid phases are given by the intersection of the tie line with the boundary of the two-liquid region. For example, at the temperature T_1, the tie line is A′−B′, and the compositions of the two liquid phases L_A and L_B are X_{A1} and X_{B1}, respectively. At the temperature T_2 corresponding to the isothermal A-B, the two liquid phases have the compositions of X_{A2} and X_{B2}, respectively, and the solid R forms by decomposition of liquid L_A ($L_A \rightarrow R + L_B$). This reaction is referred to as a monotectic reaction, and the temperature T_2 is called monotectic point. At this temperature, the number of degrees of freedom is zero as calculated from the phase rule (F = 2–3 + 1 = 0). When an intermediate compound is formed from two immiscible liquids, it is called the synthetic phase diagram.

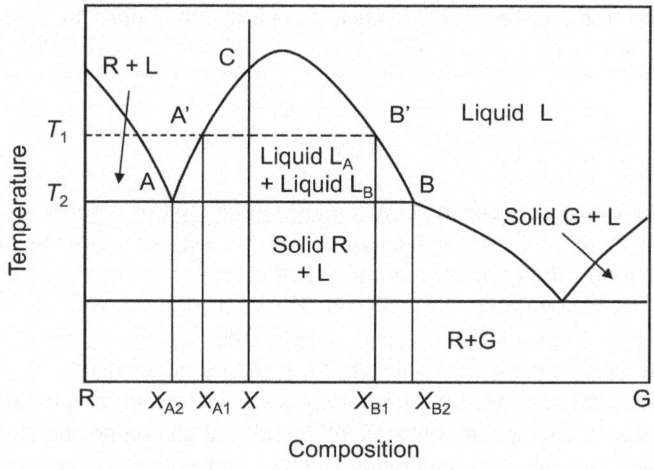

Fig. 2.13 An example of phase diagram containing immiscible liquid

2.3.2.6 Thermodynamic Considerations of Two-Component Phase Diagrams

Let us consider the two-component phase diagram thermodynamically by calculation of the free energy when atoms A and B are mixed. It is assumed that the binding energy is determined only by the nearest neighbor atoms, the internal energy is equal to the sum of the binding energy, the coordination number is Z, the total number of atoms is N, the proportion of A atoms is f_A, and the binding energies of A–A, A–B, and B–B are U_{AA}, U_{AB}, and U_{BB} ($U_{AA} > U_{BB}$), respectively. Under these conditions, the number of A and B atoms around an atom is f_AZ and $(1–f_A)Z$, respectively, and the number of A–A, B–B, and A–B interatomic bonds are $f_A{}^2ZN/2$, $(1–f_A)^2ZN/2$, and $f_A(1–f_A)ZN$, respectively.

The internal energy of the solid solution at absolute zero temperature can be expressed as

$$U = \frac{ZN}{2}\left[f_A{}^2U_{AA} + (1-f_A)^2U_{BB} + 2f_A(1-f_A)U_{AB}\right] \tag{2.7}$$

By substituting $f_A{}^2 = f_A - f_A(1 - f_A)$ and $(1 - f_A)^2 = (1 - f_A) - f_A(1 - f_A)$, the following equation can be derived:

$$U = \frac{ZN}{2}\left[f_AU_{AA} + (1-f_A)U_{BB} + 2f_A(1-f_A)\left(U_{AB} - \frac{U_{AA}+U_{BB}}{2}\right)\right] \tag{2.8}$$

The change in the enthalpy, ΔH, is equal to the sum of the change in the internal energy, ΔU, and the PV work, $\Delta(PV)$. Because $\Delta(PV)$ is much smaller than ΔU, in the solid, ΔH is almost equal to ΔU. The sum of the first and second terms in the square bracket of Eq. (2.8) is the free energy in phase separation (the dash-dot line in Fig. 2.14). The mixing entropy of the atoms can be expressed as follows:

$$S = k \ln \frac{N!}{(f_A N)![(1 - f_A)N]!} \tag{2.9}$$

where k is the Boltzmann constant. When a small amount of A (or B) is added to B (or A), the free energy decreases because of the increase in the mixing entropy. The formation of solid solution is always stable in the term of mixing entropy. When the enthalpy change is negligible, the solid solution is called the ideal solution. When the enthalpy change is considered, it is called the regular solution.

If the third term in the square bracket of Eq. (2.8) is negative (i.e., U_{AB} is smaller than $(U_{AA}+U_{BB})/2$), the free energy is totally concave up, and the solid solutions are more stable than the simple mixed state of A and B at all compositions. As a result, the relationship between the temperature and the composition is given by a complete solid-solution phase diagram (Fig. 2.14a). On the other hand, when the third term is positive in Eq. (2.8) (i.e., U_{AB} is larger than $(U_{AA}+U_{BB})/2$), there is a possibility that the free energy is concave down between the A and B compositions (Fig. 2.14b) and phase separation occurs. Sometimes, the partial solid solutions, which mean that a small amount of A or B is dissolved into B or A, respectively, are formed near A and B axes. In this case, the phase diagram becomes eutectic or peritectic. The compositions at the minimal points of the free energy near A and B axes are the solubility limits of solid solutions (α and β) as seen in Fig. 2.10. Figure 2.8 indicates no solubility near A and B axes. In this way, the phase diagram can be drawn from the dependence of the free energy on the temperature and the composition.

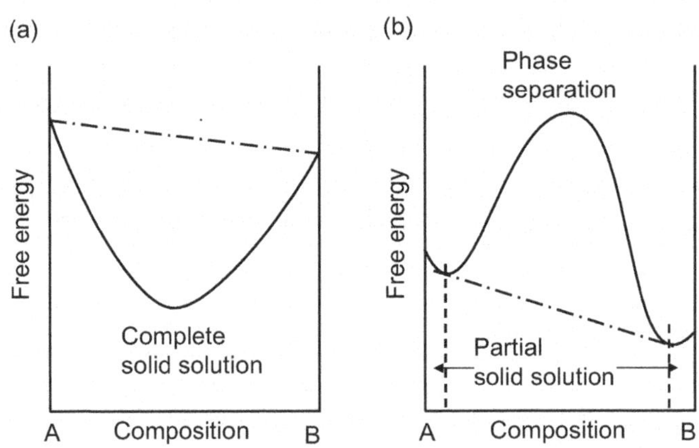

Fig. 2.14 Schematic illustration of free energy in A and B system: (**a**) complete solid solution and (**b**) phase separation

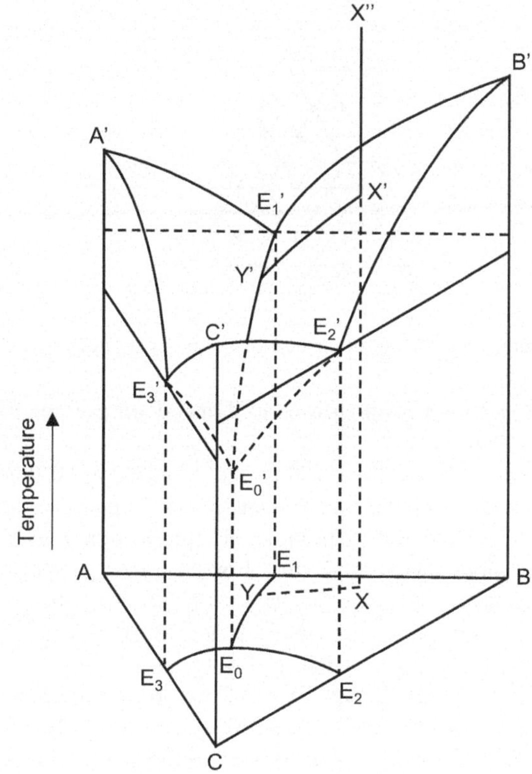

Fig. 2.15 An example of a three-component phase diagram

2.3.3 *Three-Component Phase Diagrams*

In the three-component phase diagram, the maximum number of degrees of freedom is 3 ($F = 3–P + 1 = 4–P$). Therefore, the three-component phase diagram should be drawn as a three-dimensional figure, as shown in Fig. 2.15. This is a triangular prism consisting of compositional axes on the triangle base plane and temperature axis in the perpendicular direction. As seen in the side planes, this is a combination of A–B, B–C, and C–A eutectic systems. The top of the figure indicates the liquidus surface. When the liquid of composition X″ is cooled, the composition of liquid phase migrates along liquidus surface. However, it is difficult to understand or draw such three-dimensional figures. To solve this problem, a two-dimensional triangle figure is used in which the compositions are marked on three axes of plane, and the temperature of the liquidus surface is sometimes represented as contours. Furthermore, isothermal sections, in which triangular prism is cut at a certain temperature, are also useful to understand the three-component phase diagram.

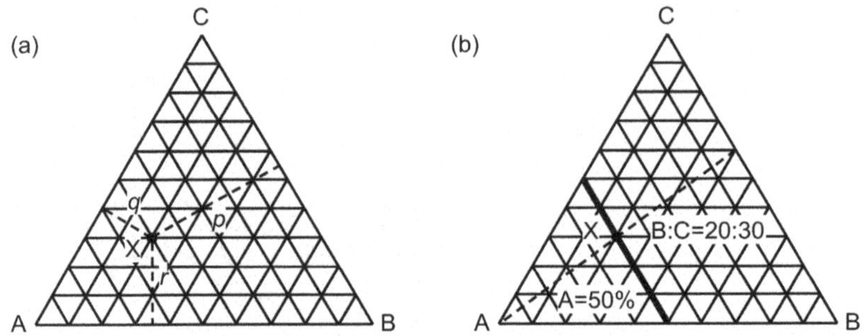

Fig. 2.16 Compositional characteristic in the triangle of three-component phase diagram

2.3.3.1 Compositions in Three-Component Phase Diagrams

The compositions in three-component phase diagrams can be read on the triangle plane as follows. (1) The lengths of the perpendicular lines drawn from the point X to the sides of the triangle are proportional to the compositions of the terminal components (Fig. 2.16a). For example, the composition of X in Fig. 2.16a is A:B:C $= p{:}q{:}r =$ 50:20:30. (2) The line drawn through the point X parallel to the opposite side of the triangle gives the proportion of two components at the fixed composition of one component. For example, the composition of X is constant as 50% A, and the compositions, B and C, are variable. Furthermore, on the line drawn thorough the point X from the corner to the opposite side of the triangle, the composition of one component is variable, and the compositional ratio of the other two components is constant. The dashed line in Fig. 2.16b indicates that B:C is equal to 2:3. The lever principle can also be used in three-component phase diagrams.

2.3.3.2 Three-Component Eutectic Phase Diagrams

An example of a three-component eutectic phase diagram is shown in Fig. 2.17, without formation of solid solutions. E_{AB}, E_{BC}, and E_{AC} are the eutectic points of two-component phase diagrams of A-B, B-C, and A-C systems, respectively. E_0 is the eutectic point of three-component phase diagram of A-B-C system. The regions separated by the connecting lines (E_{AB}-E_0, E_{BC}-E_0 and E_{AC}-E_0), indicated as A, B, and C, are called the primary phase region where primary crystal appears during cooling the melt. In other words, the solid phases coexist in equilibrium with the liquid phase in its primary phase region, which are present on the liquidus surface. Two solids and one liquid exist at equilibrium on the boundary between the primary phase regions. For example, the eutectic reaction of L $\rightleftarrows$ A + B occurs on the boundary of regions A and B (line $E_{AB} - E_0$), and L, A, and B coexist.

When a melt at the point X, as shown in Fig. 2.17, is cooled, phase B precipitates as the primary crystal at 600 °C. On further cooling along the contour line, more

Fig. 2.17 An example of a three-component eutectic phase diagram

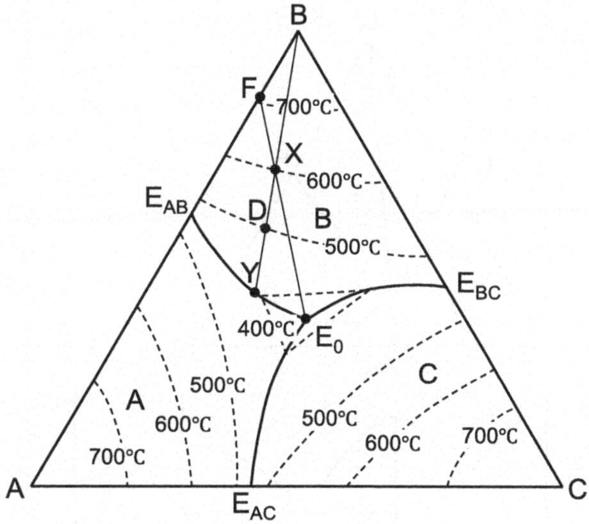

phase B forms from the liquid phase. Because the proportion of A and C in the liquid is not changed by the precipitation of B, the composition of the liquid migrates from X to Y along the line segment of B-X. The proportion of the liquid and solid phases can be calculated by the lever principle. For example, when the composition of the liquid phase is located at point D, the proportions of the solid and liquid phases are DX/DB and XB/DB, respectively. After the composition of the liquid reaches point Y, it migrates toward E_0 along the boundary line E_{AB}-E_0 where a eutectic reaction (L ⇄ A + B) takes place. Accordingly, the composition of the solid phase changes toward A. The average composition of X and the compositions of the solid and liquid phases must be present on a straight line. Therefore, when the composition of the liquid phase is at E_0, the composition of the solid phase is at F, which is the intersection of the trigonal side of A-B and the extension of E_0-X. At the point E_0, the number of degrees of freedom is 0 ($F = 3–4 + 1 = 0$), where the eutectic reaction (L ⇄ A + B + C) occurs, and the composition of solid phase migrates from F to X. The E_0 is called the ternary eutectic point.

2.3.3.3 Three-Component Phase Diagrams with Alkemade Line

Intermediate compounds of two or more components also form in a three-component system in the same way as in a two-component phase diagram. In this case, a primary phase region of a compound is separated by boundaries with the other phases. When the primary phase regions of two phases have a common boundary line, a straight line connecting the compositions of the two phases is drawn as a so-called Alkemade line. It cannot be drawn between phases without a common boundary line. The Alkemade line divides a ternary phase diagram into small compositional triangles.

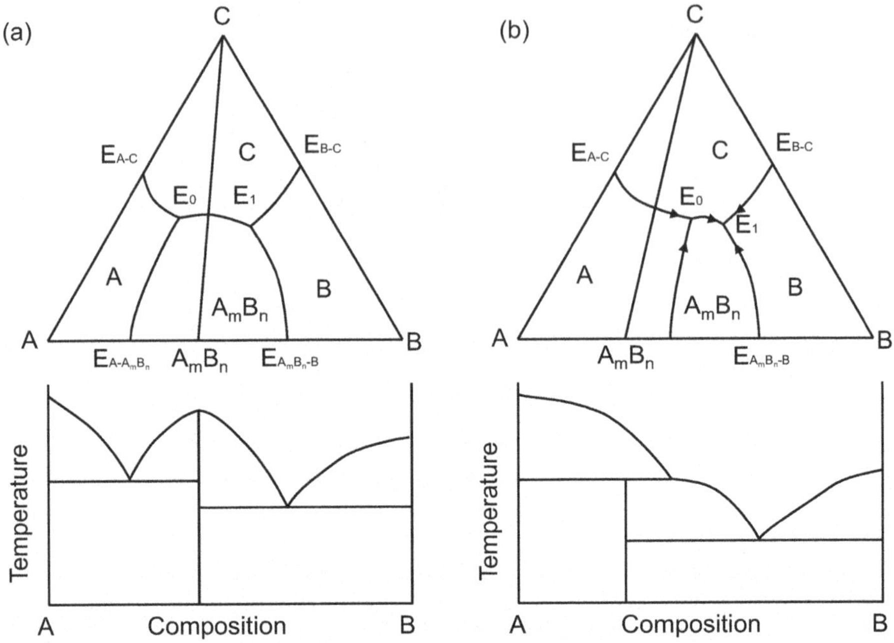

Fig. 2.18 Examples of three-component phase diagrams with Alkemade line

Figure 2.18 shows typical three-component phase diagrams in A-B-C system including an intermediate compound. The intermediate compound A_mB_n is formed between A and B. The connecting line between A_mB_n and C is the Alkemade line. In this diagram, triangles A-C-A_mB_n and B-C-A_mB_n are the divided compositional triangles. The triangles are characterized by invariant points, such as eutectic point or peritectic point. Final average composition of solid phase agrees with the initial composition of liquid phase, and final composition of liquid phase is equal to the invariant point.

As shown in Fig. 2.18a, when the boundary line (E_0-E_1) intersects with the corresponding Alkemade line (C-A_mB_n), the phase diagram consists of two three-component eutectic phase diagrams of A-C-A_mBn and B-C-A_nB_m systems. This is understood from schematically illustrated cross-sectional image. A_mB_n congruently melts. E_0 and E_1 are ternary eutectic points. The temperature is the lowest on the Alkamade line and the highest on the boundary line at the intersection of E_0-E_1 and C-A_mB_n. When the initial composition of liquid phase (i.e., the average composition in the system) is left or right from Alkemade line, the composition of liquid phase migrates toward E_0 or E_1, respectively, during cooling, giving the final composition of liquid phase. At these points, the eutectic reactions take place as $L \rightleftarrows A + C + A_mB_n$ at E_0 and $L \rightleftarrows B + C + A_mB_n$ at E_1.

On the other hand, when the Alkemade line and the boundary line do not cross, the phase diagram is peritectic. Since A_mB_n is apart from the primary phase region, the phase diagram becomes peritectic. Figure 2.18b is the combination of peritectic and eutectic phase diagrams. A_mB_n incongruently melts, that is, decomposition melting takes place. In this case, E_0 is ternary peritectic point, and E_1 is ternary eutectic point. Arrows indicate the direction of the compositional change of the liquid phase during cooling. When the composition of the initial liquid phase is present in the triangle A-C-A_mB_n, the composition of liquid phase migrates to E_0 during cooling, where the peritectic reaction ($L + A \rightleftarrows C + A_mB_n$) occurs. Within the triangle B-C-A_mB_n, the composition of liquid phase migrates to E_1, where the eutectic reaction ($L \rightleftarrows B + C + A_mB_n$) occurs. Although the actual phase diagram has many intermediate compounds and is very complex, it is possible to decipher it according to the basic concepts stated above.

References

1. W. D. Kingery, *Introduction to Ceramics,* 2nd edn. (Wiley, New York, NY, 1975) p. 247
2. P. Atkins, J. de Paula, J. Keeler, *Atkins' Physical Chemistry,* 11th edn. (Oxford University Press, Oxford, 2017) p. 119
3. C. G. Bergeron, S. H. Risbud, *Introduction to Phase Equilibria in Ceramics* (Westervill, OH. 1984)
4. A. M. Alper, *Phase Diagram in Advanced Ceramics* (Elsevier, Amsterdam, 1995)
5. E. M. Levin, C. R. Robbins, H. F. McMurdie, *Phase Diagrams for Ceramics*, vol. **I**, (1964)
6. J.W. Gibbs, *Trans. Conn. Acad.* **III**, 108 (1876)

Chapter 3
Solid State Reactions and Sintering

Satoshi Tanaka

Abstract Solid state reactions are mainly used to synthesize fine raw particles and to sinter advanced ceramics. To control the powder characteristics and sintering process, several analyses have been conducted based on classical thermodynamics and kinetics. In this chapter, we describe the thermodynamics and kinetics including the temperature-dependent standard free energies of formation and diffusion coefficients. Then, practical studies of solid state reaction kinetics are introduced using classical theories, such as the Jander and Ginstling-Brounstein equations. Finally, we discuss classical sintering theory using a two-particle model. Although several novel sintering methods recently have been developed, classical sintering theory is still important.

Keywords Solid state reaction · Thermodynamics · Diffusion · Powder synthesis · Sintering

3.1 Features of Solid State Reactions

Solid state reactions are classified as either reactions involving a change in chemical composition—such as the reactions between the solid and vapor, solid and liquid, and solid and solid phases—or those not involving a change in chemical composition, such as sintering and phase transitions. Reaction rates have been proposed for solid state reactions between single crystals and particles. It is well known that vapor and liquid phase reactions proceed homogeneously owing to the homogeneous distributions of the components independent of a place and time. Reaction rates can be described as functions of concentration and time. In contrast, solid state reactions between solid-solid, solid-liquid, and solid-vapor phases only occur at contact interfaces between two solids and progresses by mass diffusion over phase

S. Tanaka (✉)
Department of Materials Science and Technology, Nagaoka University of Technology, Nagaoka, Japan
e-mail: stanaka@vos.nagaokaut.ac.jp

J. Hojo (ed.), *Materials Chemistry of Ceramics*,
https://doi.org/10.1007/978-981-13-9935-0_3

boundaries, so the overall reaction rates are determined by the mobility of the boundaries between reactants and products. In particular, the atomic-level mixing of reactants is impossible in the solid-solid reactions, and the reactions start from the particle contact points, so particle sizes significantly affect reaction rates. For sintering and phase transition, however, mass transfers occur in closed ranges, so the experimental data have been treated theoretically for the diffusion mechanisms.

As previously mentioned, solid state reactions depend on material characteristics including not only particle sizes and specific surface areas but also point defects, crystal lattice irregularities, and crystallinities. Because such intrinsic parameters affect the kinetics of solid state reactions, they are often controlled to improve reaction efficiency, to reduce reaction temperature, and to accelerate sintering rates. For example, low-crystallinity nanoparticles increase solid state reaction rate and reduce reaction temperature.

Furthermore, extrinsic parameters such as pressure, atmosphere, electric and magnetic fields, radial irradiation, and phase transfer affect solid reactivity, and controlling such parameters is applicable to activate solid state reactions, reducing reaction temperature and controlling product morphologies and shapes. For crystallographic transformations of solids, the transition states effectively enhance solid reactivity, a phenomenon referred to as the Hedvall effect [1].

3.2 Thermodynamics of Solid State Reactions [2–4]

3.2.1 Activation Energies

Although the actual solid state reaction processes are very complicated, their fundamental processes and equilibria are based on chemical thermodynamics, so the basic mechanism of solid state reactions is mutual atomic diffusion, which is similar to the usual mechanism of other atomic-level chemical reactions. However, solid state reactions only occur at material boundaries, and the reactions proceed heterogeneously. For solid state reactions to progress, the reactant atoms must come to the boundaries and then diffuse from there throughout the whole material matrix. Further, the reactants must temporarily pass through activation states (i.e., activated complexes). The activation energy, ΔG^*, is the energy barrier that must be surmounted for the reaction to progress and is defined as the difference between the standard free energies of formation of the reactants and the energy of the activated complex. The reaction rate constant, k, is expressed in terms of activation energy as follows:

$$k = k_0 \exp\left(-\frac{\Delta G^*}{RT}\right) = k_0 \exp\left(\frac{\Delta S}{R}\right) \exp\left(-\frac{\Delta H^*}{RT}\right), \tag{3.1}$$

where ΔS and ΔH are the entropy and enthalpy of the activated complex, respectively. If a solid state reaction consists of multiple elementary reactions, each reaction will satisfy Eq. (3.1). The total reaction rate is limited by the reaction showing the highest activation energy, ΔG^*, which is referred to as the rate-determining process. For simultaneous reaction pathways, the reactions will progress through the pathway showing the lowest activation energy, ΔG^*, which is estimated from the temperature dependence of the reaction rate constant as follows:

$$\frac{d\ln k}{dT} = \frac{\Delta G^*}{RT^2}, \tag{3.2}$$

All solid state reactions consist of diffusion and chemical reaction. Diffusion in the solid limits the rates of solid state reactions since chemical reactions usually occur faster than diffusions. Further, other physical chemistry factors such as phase transition, crystallization, sintering, sublimation, and melting either independently or simultaneously affect solid state reactions. The total rates and mechanisms of solid state reactions are affected by the physical chemistry factors of all the individual elemental process. The reaction rate constant, $\bar{k}$, in the total reaction rate is expressed similar to those in the rates of other chemical reactions as follows:

$$\bar{k} = \bar{k}_0 \exp\left(-\frac{Q}{RT}\right), \tag{3.3}$$

where Q is the apparent activation energy and depends on the activation enthalpies of diffusion, chemical reaction, nucleation, and other limiting processes. Reaction rates are often increased by the small amounts of additives or by the contributions of liquid phases because the additives either leads to low-ΔG^* elementary processes or behaves as catalysts thereby reducing ΔG^*. Controlling activation energies is important for synthesizing materials.

3.2.2 *Standard Free Energies of Formation*

The standard free energies of formation of solid state reactions, ΔG_r°, are considered the same as those of general chemical reactions. ΔG_r° is defined as the difference between the standard free energy of the reactant and product. Ellingham diagrams shown in Fig. 3.1 [4] are well known and show ΔG_r° for 1 mol oxide materials at constant temperature and pressure. The general thermodynamics equation is $\Delta G^\circ = \Delta H^\circ - T\Delta S^\circ$. Since ΔH° and ΔS° are temperature-independent constants, ΔG° monotonically changes with increasing temperature along a slope of $-\Delta S^\circ$ (Fig. 3.2). For oxidation, ΔG° increases with increasing temperature because the entropy change by consumption of gas phase is negatively large, $\Delta S^\circ < 0$. ΔG° is independent of temperature and is constant for the reaction, $C + O_2 \rightarrow CO_2$, because

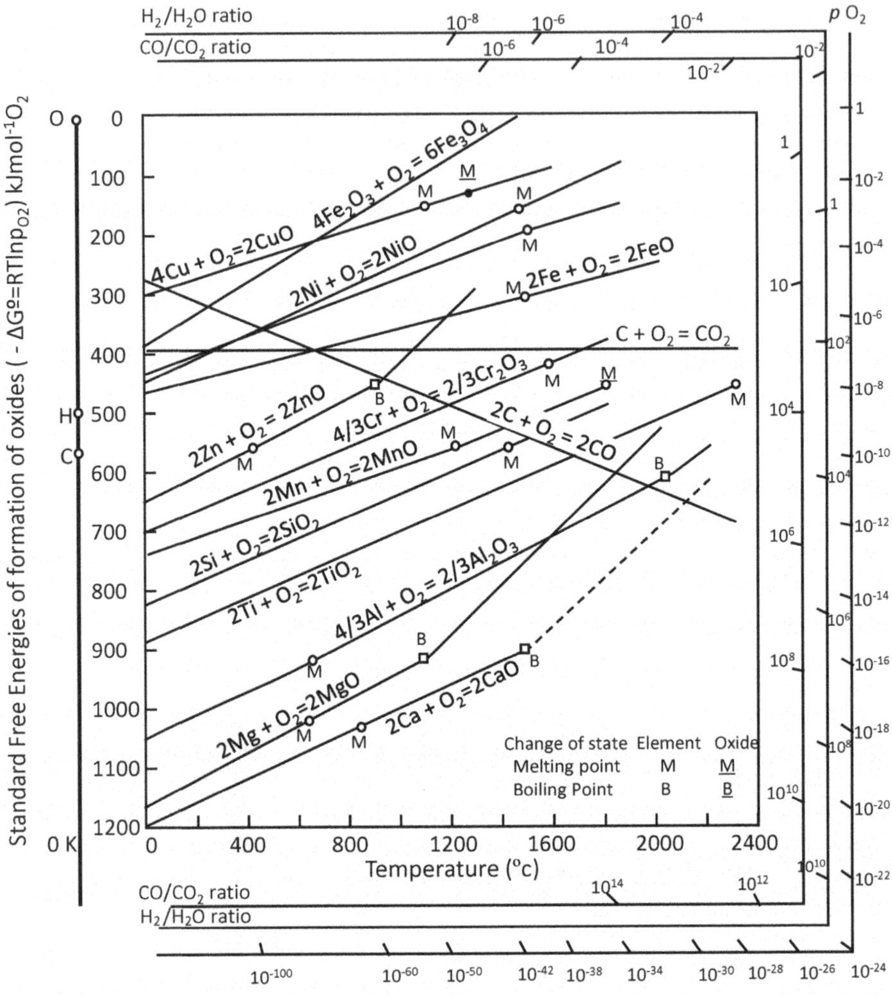

Fig. 3.1 The standard free energy of formation of oxides as a function of temperature at the edges scales in CO/CO_2 ratio, H_2/H_2O ratio, and pO_2. From Ref. [4].

of no gas volume change, $\Delta S^\circ = 0$. In contrast, ΔG° decreases with increasing temperature for the reaction, $2C + O_2 \rightarrow 2CO$, because of gas volume expansion, $\Delta S^\circ > 0$. If all the reactants in solid state reactions are solids, the change in volume and ΔS° are both approximately zero, that is, $\Delta G_r^\circ = \Delta H_r^\circ$. When $\Delta G_r^\circ < 0$, the reaction will progress, meaning that the standard free energy of the solid state reaction should be $\Delta G_r^\circ = \Delta H_r^\circ < 0$ and that the reaction from solid to solid phase is exothermic, as referred to the rule of Van't Hoff. Furthermore, ΔG_r° can be estimated from the difference between the lattice energies of the reactants and the products.

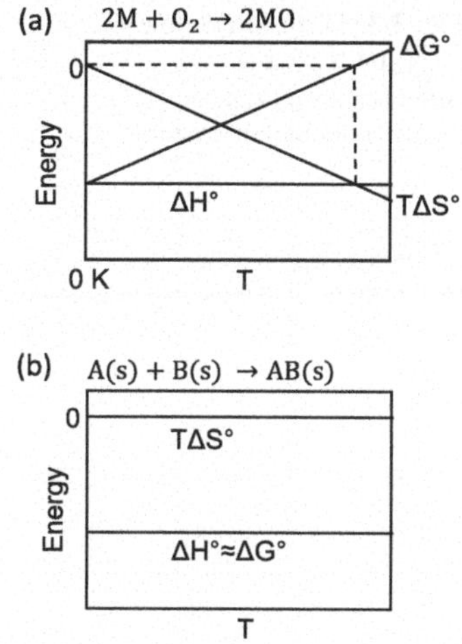

Fig. 3.2 Temperature dependence of free energy change for solid state reaction, (**a**) oxidation reaction, (**b**) solid state reaction

3.2.3 *Equilibria of Solid State Reactions*

The equilibria of metal oxide syntheses (e.g., $M + O_2 \rightarrow MO_2$) or metal oxidations depend on the oxygen partial pressure, where each activity of metal or metal oxide is assumed to be 1, and the equilibrium constant, *K*, can be expressed as follows:

$$K = \frac{1}{P_{O_2}^{eq}}, \tag{3.4}$$

where ${P_{O2}}^{eq}$ is the equilibrium oxygen pressure. When a reaction reaches thermodynamic equilibrium, the motivating energy to progress the reaction is 0, so

$$\Delta G = \Delta G^{\circ} - RT \ln P^{\circ}_{O_2} = 0, \tag{3.5}$$

where $RT\ln P^{\circ}{}_{O2}$ means the difference between the free energies depending on the oxygen partial pressure. According to Eq. (3.5), if ΔG° is measured at a certain temperature from the Ellingham diagram, the equilibrium oxygen partial pressure can be calculated at any temperature. For only solid state reactions, if the reaction energy is less than 0, i.e., $\Delta G_r{}^{\circ} < 0$, the reaction apparently finished at equilibrium.

3.2.4 *Diffusion Couples and Phase Diagrams [5]*

When solid state reactions of A and B phases, for example, occurs in eutectic systems, the α+β eutectic crystal phase appears at the interphase boundaries between A and B diffusion couples, as shown in Fig. 3.3. In practice, ionic radii and balances are slightly different, and the partial solid solution, α or β phase, appears near the

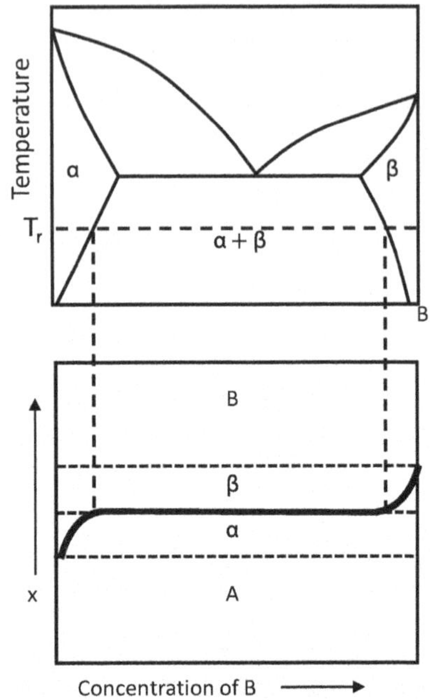

Fig. 3.3 Diffusion couple of A and B phases corresponding to phase diagram in eutectic system

interfaces. The interfaces between A and B diffusion couples are the boundaries of the solid solution α and β phases. When new compositions of AB form by the reactions of this diffusion couple, the interfaces between A and AB consist of α and γ, which are partial solid solution of the A and AB phases, respectively, while the interface of AB and B consists of γ and β, which are AB phase and partial solid solution of B, respectively. The reaction progresses forming of the AB product phase.

3.3 Practical Studies of Solid State Reaction Kinetics

3.3.1 Experimental Determination of Solid State Reaction Kinetics

The kinetics of the solid state reactions are examined by various experimental methods. Thermogravimetric analysis (TGA) measures the changes in weight of materials accompanied by the reaction as a function of temperature and time. Differential thermal analysis (DTA) gives the endothermic or exothermic temperature change, and differential scanning calorimetry (DSC) gives the calorific change as a function of temperature. The crystal phases or compositions of the reactants and products are examined by powder X-ray diffraction (XRD). TG-DTA, TG-DSC, and TG-XRD are often used simultaneously for the kinetic analysis of the reaction. Further, scanning electron microscope (SEM) and transmission electron microscopy

(TEM) with energy dispersive X-ray spectrometry (EDX) are used to characterize nanoparticle morphologies and elemental contents. In addition, gas chromatography (GC), Raman spectrophotometry, infrared (IR) absorption spectrometry, and dielectric and magnetic measurements are utilized. The atomic-level reactivities of ions in solid state coordination environments can be quantitatively monitored using magic angle spinning-nuclear magnetic resonance (MAS-NMR) spectroscopy for solid ^{27}Al and ^{29}Si. These analytical methods are utilized to determine reaction rate equations and compare parameters, to compare the activation energies obtained from the temperature-dependent parameters in the Arrhenius plots, to measure reaction rates (i.e., the rates at which products are formed by solid state reactions) under constant temperatures and ambient pressures for various reaction times, to compare the initial and final reaction temperatures obtained from TG-DTA curves for constant high-temperature reactions, to compare the residual reactant and product phases detected from IR and XRD spectra, and to determine sintering rates by measuring shrinkage and grain growth rates.

Differential thermal analysis (DTA) is used to measure the thermal energies of solid state reactions. When reactants are heated at a constant rate, the reaction heat, which is endothermic or exothermic, is measured from the difference between the temperatures of the reactants and the standard material (i.e., α-alumina powder). We can obtain information about the thermodynamics of endothermic and exothermic reactions from DTA plots as functions of temperature and from the changes in temperature from the start to the end of reactions. However, since the crystal phases and chemical species of reaction products cannot be determined from DTA plots alone, they are simultaneously identified using XRD, IR, Raman, spectroscopies, and gas analysis.

For example, Fig. 3.4 shows DTA plots for various kaolinite powders [6]. Although kaolinite compositions depend on the mineral natures, the following series of reactions generally occurs and corresponds with the DTA plots shown in Fig. 3.4. At ~600 °C, an endothermic reaction is caused by dehydration. Then, in the range 800–1000 °C, metakaolin is formed, or alumina phase transition occurs by an exothermic reaction. The solid solution with mullite appears above 1100 °C. Figure 3.4 indicates that the reaction heats and temperature in the DTA plots vary depending on the composition of minerals.

3.3.2 *Powder Reactions [7–9]*

3.3.2.1 Features of Powder Reactions

Powder synthesis is a representative solid state reaction. The reaction between raw particles starts at the surfaces of contacted particles, where product layers form with increasing reaction time. Then, the reactant particles and the product layers sequentially react at the interfaces between particles and layers, and the interfaces move to the non-reacted particles. Either the interfacial chemical reaction rate or the diffusion rate limits the overall reaction rate. The diffusion depends on the ionic species in the

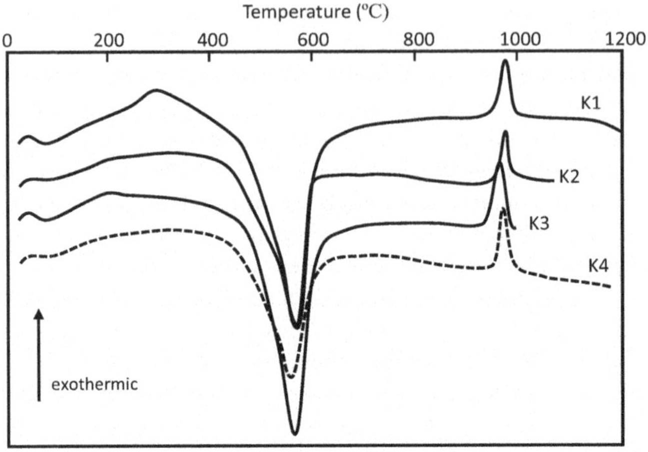

Fig. 3.4 Differential thermal analysis of kaolins during heating: K1, kaolinite and muscovite; K2, kaolinite, muscovite, and quartz; K3, kaolinite, illite, and quartz; K4, kaolinite and quartz. (From ref. [6]. It is partially modified)

crystal lattice, the ionic concentration gradient, the vacancy concentration, the temperature, etc. If the ionic diffusion is slower than the chemical reaction, the diffusion limits the overall reaction.

3.3.2.2 Diffusion-Controlled Reaction Rates

Equations for powder synthesis reaction rates have been proposed to control powder characteristics. Among them, the Jander equation is simple and versatile, which is based on the illustration in Fig. 3.5a. For the Jander equation, some conditions are assumed for simplicity: (i) diffusion limits the reaction, (ii) ions diffuse unidirectionally, (iii) the surface reactions occur quickly and the product homogeneously covers the particle surface, (iv) the particle size does not change, (v) the volume ratio of the reactant and product is 1:1, and (vi) the diffusion coefficient is a time-independent constant. Here, the non-reacted particle volume, V, of a spherical particle is as follows:

$$V = \frac{4\pi(r_0 - x)^3}{3}, \tag{3.6}$$

where x is the thickness of the reacted phase, i.e., the product layers, and r_0 is the particle radius. V also can be expressed using the reaction ratio, α, as follow:

$$V = \frac{4\pi r_0^3}{3}(1 - \alpha), \tag{3.7}$$

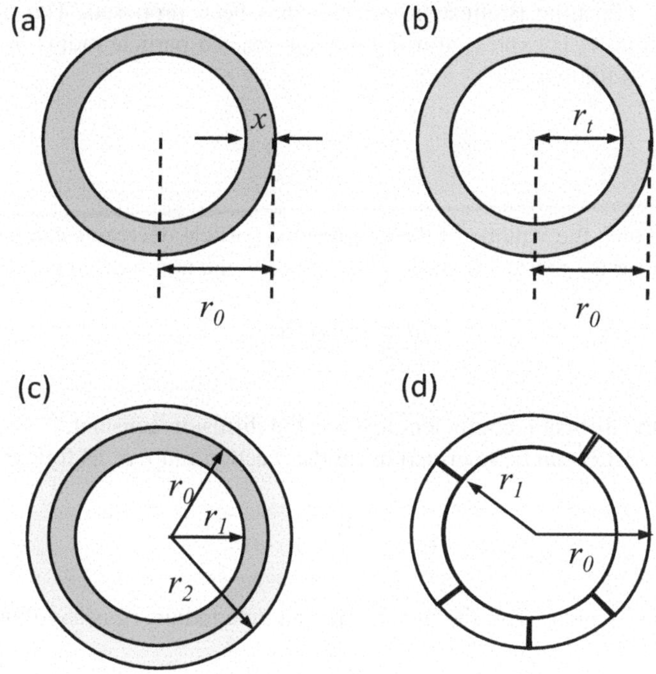

Fig. 3.5 Schematic illustration of model particle for each reaction equation. (**a**) Jander equation, (**b**) Ginstling-Brounstein equation, (**c**) Carter equation, (**d**) rate-controlled interface reaction

Since Eqs. (3.6) and (3.7) are identical, the thickness of the product layer, x, can be expressed as follows:

$$x = r_0\left[1 - (1 - \alpha)^{1/3}\right], \tag{3.8}$$

where x is proportional to the square root of time, and the reaction proceeds according to the parabolic rate law as follows:

$$x^2 = kt, \tag{3.9}$$

$$\alpha^2 = kt. \tag{3.10}$$

The Jander equation is then expressed as follows:

$$\frac{kt}{{r_0}^2} = \left[1 - (1 - \alpha)^{1/3}\right]^2. \tag{3.11}$$

Similarly, Ginstling-Brounstein equation has been proposed. The non-reacted particle volume, V, is expressed using the non-reacted particle radius, r_t, as shown in Fig. 3.5b, as follows:

$$V_t = \frac{4\pi r_t^{3}}{3}. \tag{3.12}$$

The rate at which the volume of the non-reacted particle decreases can be related to the equation for the thermal diffusion flow through the non-reacted phase as follow:

$$-\frac{dV_t}{dt} = \frac{4\pi kDr_0 r_t}{r_0 - r_t}, \tag{3.13}$$

where D is the diffusion coefficient and k is the diffusion constant.

Equation (3.12) can be rewritten using the reaction ratio, α, as follows:

$$V_t = \frac{4\pi r_0^{3}}{3}(1-\alpha). \tag{3.14}$$

From Eqs. (3.7) and (3.13), the non-reacted particle radius, r_t, is as follows:

$$r_t = r_0(1-\alpha)^{1/3}. \tag{3.15}$$

Differentiating Eq. (3.12) yields

$$-\frac{dV_t}{dt} = -4\pi r_t^{2}\frac{dr}{dt}. \tag{3.16}$$

Since Eqs. (3.13) and (3.16) are identical, the differential equation is derived,

$$r_t dr_t - \frac{r_t^{2}}{r_0}dr_t = -kDdt. \tag{3.17}$$

Integrating Eq. (3.17) and using the boundary conditions $r_t = r_0$ at $t = 0$ yields

$$\frac{r_t^{2}}{2} - \frac{r_t^{3}}{3r_0} = -kDt + \frac{r_0^{2}}{2}. \tag{3.18}$$

By substituting Eq. (3.15) into Eq. (3.18), the Ginstling-Brounstein equation is obtained.

$$k_{G.B}t = \frac{2kDt}{r_0} = 1 - \frac{2\alpha}{3} - (1-\alpha)^{2/3}. \tag{3.19}$$

The Jander equation, Eq. (3.11), and the Ginstling-Brounstein equation, Eq. (3.19), are both based on the assumption that the molar volume of the product phase is equal to that of the non-reacted phase. If this condition is not applicable to the reaction, then the relation between the product and the non-reacted phase is modified using a novel parameter, Z, on the illustration in Fig. 3.5c. Since the total volume consists of those of the non-reacted and product phases,

$$\frac{4\pi r_2{}^3}{3} = \frac{4\pi(r_0{}^3 - r_t{}^3)Z}{3} + \frac{4\pi r_t{}^3}{3}, \tag{3.20}$$

where r_0 is the initial particle radius, r_t is the radius of the non-reacted particle at time t, and r_2 is the total particle radius, i.e., the sum of the radii of the non-reacted and product phases at time t. Equation (3.20) can be simplified as follows:

$$r_2{}^3 = Zr_0{}^3 + r_t{}^3(1 - Z). \tag{3.21}$$

Using the Ginstling-Brounstein equation, Eq. (3.21) can be expressed according to the Carter equation as follows:

$$\frac{2ktD}{r_0{}^2} = \frac{Z - (Z-1)(1-\alpha)^{2/3} - [1 + (Z-1)\alpha]^{2/3}}{Z - 1}. \tag{3.22}$$

3.3.2.3 Rate-Controlled Interface Reactions

When diffusion rates are sufficiently high, interfacial reactions control the overall reaction rates, whose equations are derived for the interfacial-reaction-controlled reaction rates on the illustration in Fig. 3.5d. For the derivation, the following assumptions are made: (i) the interfacial reaction at the phase boundary limits the reaction rate, (ii) the reaction rate depends on the surface area of the non-reacted phase, and (iii) nucleation instantly occurs and the nuclei cover the non-reacted phase.

The reaction progresses proportionally to the surface area of the non-reacted phase as follows:

$$-\frac{dV_t}{dt} = kS_t, \tag{3.23}$$

where V_t and S_t are the volume and surface area of the non-reacted phase, respectively, at time t. Equation (3.23) then becomes

$$-\frac{d(4\pi r_t{}^3/3)}{dt} = k4\pi r_t{}^2, \tag{3.24}$$

and

$$-\frac{dr_t}{dt} = k. \tag{3.25}$$

Thus, the reaction rate is a constant corresponding to the rate at which the radius of the non-reacted layer of the particle decreases. The ratio of the non-reacted volume $(1-\alpha)$ is expressed as follows:

$$1 - \alpha = \frac{V_t}{V_0} = \frac{r_t^3}{r_0^3}, \tag{3.26}$$

where, V_0 is the initial particle volume. Equation (3.26) can thus be expressed as follows:

$$r_t^2 = (1 - \alpha)^{2/3} r_0^2. \tag{3.27}$$

The reaction rate is derived from Eqs. (3.25), (3.26), and (3.27) as follows:

$$-\frac{d(1-\alpha)}{dt} = \frac{d(V_t/V_0)}{dt} = \frac{kS_t}{V_0}. \tag{3.28}$$

Then,

$$\frac{d\alpha}{dt} = -\frac{k4\pi r_t^2}{4\pi r_0^3/3} = \frac{3k(1-\alpha)^{2/3}}{r_0}, \tag{3.29}$$

so the reaction rate can be expressed as follows:

$$1 - (1-\alpha)^{1/3} = \frac{K't}{r_0}, \tag{3.30}$$

which is called the contraction equation, and it is used for interfacial-reaction-controlled reaction rates and the hydrogen reduction of metal oxide particle.

3.3.2.4 Nucleation and Growth

For homogeneous particle phase transitions, the kinetics are expressed by the Avrami-Erofeev equation:

$$\frac{V_t}{V_0} = 1 - \exp[-(kt)^n], \tag{3.31}$$

where V_0 and V_t are the initial particle volume and the volume of the reacted phase at reaction time t, respectively, and n is a constant depending on the reaction mechanism, nucleation rate, and nuclei geometries.

3.3.2.5 First-Order Reaction Equations

Although the order of solid state reactions do not have any physical meaning, the rates of nucleation may be expressed as the first-order reactions as follows:

$$\ln(1-\alpha) = -kt. \tag{3.32}$$

3.3.3 Analysis of Reaction Rates [10]

Six equations proposed in Section 3.3.2 for the rates of powder reactions, i.e., Eqs. (3.10), (3.11), (3.19), (3.22), (3.30), and (3.32), can be rearranged as follows:

$$D_1(\alpha) = \alpha^2 = kt. \tag{3.33}$$

$$D_3(\alpha) = \left[1-(1-\alpha)^{1/3}\right]^2 = \frac{kt}{r_0{}^2},. \tag{3.34}$$

$$D_4(\alpha) = \frac{2kDt}{r_0} = 1 - \frac{2\alpha}{3} - (1-\alpha)^{2/3} = k_B t,. \tag{3.35}$$

$$R_3(\alpha) = 1-(1-\alpha)^{1/3} = \frac{k_t t}{r_0},. \tag{3.36}$$

$$A_3(\alpha) = \sqrt[3]{-\ln(1-\alpha)} = kt,. \tag{3.37}$$

$$F_1(\alpha) = \ln(1-\alpha) = -kt,. \tag{3.38}$$

All these equations are functions of the reaction time. Here, when the reaction ratio, α, is 0.5, time is taken as $t_{0.5}$. For example, using Eq. (3.38),

$$F_1(0.5) = \ln(1-0.5) = -kt_{0.5}. \tag{3.39}$$

The ratio $t/t_{0.5}$ is defined as the conversion time, and Eq. (3.38) is then divided by Eq. (3.39) as follows:

$$\frac{F(\alpha)}{F_1(0.5)} = \frac{t}{t_{0.5}}, \text{ and } F(\alpha) = F_1(0.5)\frac{t}{t_{0.5}}. \tag{3.40}$$

When the relation between α and $(t/t_{0.5})$ is plotted for various values α, a curve is obtained, as shown in Fig. 3.6. Using the other equations, relations between α and $(t/t_{0.5})$ can be obtained.

For the experimental data of solid state reactions, the relationship between α and $(t/t_{0.5})$ is calculated using the reaction ratio, α, and reaction time, t. The reaction mechanism can be examined by comparing the experimentally obtained data, α and $(t/t_{0.5})$, and the curves, as shown in Fig. 3.6. In practical reactions, e.g., oxidation

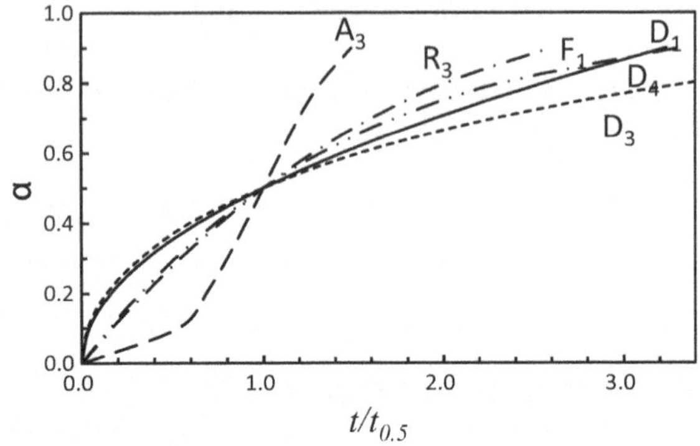

Fig. 3.6 Plots of fraction reacted, α, vs. $t/t_{0.5}$ calculated for various solid state reaction equations. (It is modified from Ref. [10]

reactions, the rate-determining step may change as the reaction proceeds. The various reaction mechanisms can be estimated from the comparison using the reaction rate equation.

3.3.4 Bulk Solid Reactions [11]

Let us consider bulk solid reactions such as oxidation and addition. In these reactions, the solid reacts with gas, liquid, or solid phases to form a surface product. Then, the product layer thickens by the diffusion of molecule, atom, or ion through the reactant phase, for example, as in Si oxidation.

Figure 3.7 shows a three-step model of Si wafer oxidation, which occurs as follows:

(i) O_2 is transferred to a SiO_2 surface where it adsorbs.
(ii) The O_2 (or O^{2-}) diffuses across the SiO_2 phase to the interface of Si and SiO_2 phases.
(iii) The SiO_2 phase is produced at the Si/SiO_2 interface by the chemical reaction of Si and O_2 (or O^{2-}).

Here, flux (mol m^{-2} s^{-1}) is defined as the transfer of substance per unit area (m^2) per in unit time (s).

(i) Under steady-state diffusion and chemical reactions, the O_2 flux from the gaseous phase to the SiO_2 surface of the Si wafer is expressed as follows:

$$F_1 = h(C^* - C_0), \quad (3.41)$$

where h is the transfer coefficient of O_2, $C*$ is the concentration of O_2 exterior to the SiO_2 surface, and C_O is the equilibrium concentration of O_2 at the SiO_2 surface. C_O is proportional as the oxygen partial pressure according to the Henry law.

Fig. 3.7 Oxidation model of Si wafer

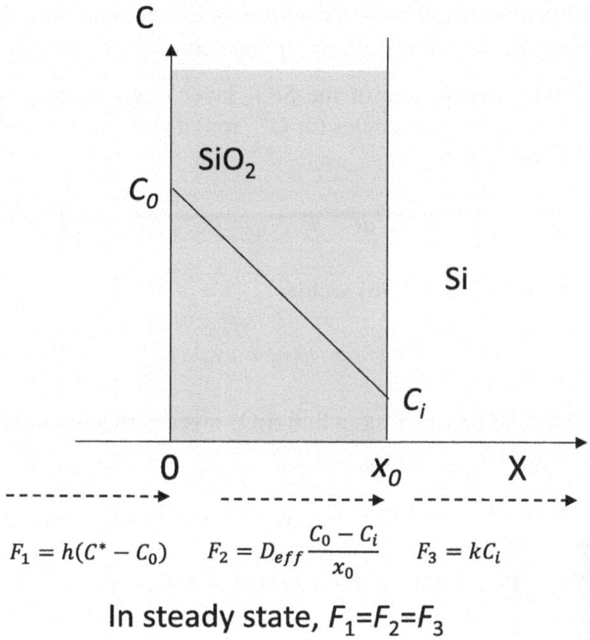

$$C_0 = Kp. \tag{3.42}$$

(ii) The flux of O_2 (or O^{2-}) during the diffusion across the SiO_2 phase is expressed as follows:

$$F_2 = -D_{eff}(dC/dx) = D_{eff}(C_0 - C_i)/x_0, \tag{3.43}$$

where D_{eff} is the effective diffusion coefficient, x_0 is the thickness of the SiO_2 layer, and C_i is the concentration of O_2 (or O^{2-}) at the interface between the SiO_2 and the Si phases.

(iii) Si oxidation at the interface is assumed to be a first-order reaction:

$$F_3 = kC_i, \tag{3.44}$$

where k is the reaction rate constant.

The oxidation is assumed to progress through the following three steps under steady-state, and $F_1 = F_2 = F_3$, so the following equation is derived:

$$\begin{aligned} F_1 = F_2 = F_3 &= kC^*/(1 + k/h + kx_0/D_{eff}) \\ &= C^*/(1/k + 1/h + x_0/D_{eff}). \end{aligned} \tag{3.45}$$

If the adsorption determines the reaction rate, $h << k$, D_{eff}, and $F = hC*$.

If the chemical reaction determines the reaction rate, $k << h$, D_{eff}, and $F = kC*$.
If the diffusion determines the reaction rate, $D_{eff} << h$, k, and $F = C*D_{eff}/x$.

The growth rate of the SiO_2 layer is given by $dx/dt = F/N_1$ (where N_1 is the number of O_2 molecules (or O^{2-} ion) diffused into a unit volume of the SiO_2 phase). Therefore,

$$dx/dt = F/N_1 = kC^*/\left[N_1\left(1 + k/h + kx_0/D_{eff}\right)\right]. \tag{3.46}$$

Integrating Eq. (3.46) yields

$$(1 + k/h)x_0 + kx_0{}^2/D_{eff} = (kC^*/N_1)t + const. \tag{3.47}$$

Since the Si wafer has a thin SiO_2 layer with a thickness of x_i at the initial condition $t = t_O$,

$$const. = (1 + k/h)x_i + kx_i{}^2/D_{eff}. \tag{3.48}$$

Thus, Eq. (3.47) can be rewritten as follows:

$$\begin{aligned} &x_0{}^2 + Ax_0 = Bt + x_i{}^2 + Ax_i \\ &\mathrm{A} = 2D_{eff}/(1/k + 1/h), \quad B = 2D_{eff}C^*/N_1, \end{aligned} \tag{3.49}$$

which can be rearranged as follows:

$$\begin{aligned} &x_0{}^2 + Ax_0 = B(t + \tau) \\ &\tau = (x_i{}^2 + Ax_i)/B. \end{aligned} \tag{3.50}$$

The thickness can be expressed as a function of time as follows:

$$x_0/(A/2) = \left[1 + (t + \tau)/\left(A^2/4B\right)\right]^{1/2} - 1. \tag{3.51}$$

When $x >> A^2/4B$ and, $t >> \tau$,

$$x_0/(A/2) \cong \left[t/\left(A^2/4B\right)\right]^{1/2} \tag{3.52}$$

or,

$$x_0{}^2 \cong Bt, \tag{3.53}$$

which is a parabolic law, where B is the reaction rate constant.

When $t << \tau$,

$$x_0/(A/2) \cong (1/2)(t+\tau)/(A^2/4B) \tag{3.54}$$

or,

$$x_0 \cong (B/A)(t+\tau), \tag{3.55}$$

which is a linear rate law, where *B*/*A* is the reaction rate constant containing *h* and *k*.

$$B/A = (C^*/N_1)/(1/h + 1/k). \tag{3.56}$$

If $h >> k$, the interfacial chemical reaction is the rate-determining step given by

$$B/A = k(C^*/N_1), \tag{3.57}$$

and if $k >> h$, the O_2 transfer is the rate-determining step given by

$$B/A = h(C^*/N_1). \tag{3.58}$$

Figure 3.8 shows the thicknesses change of SiO_2 layers on Si wafers oxidized in a dry environment at various temperatures. Figure 3.9 shows SiO_2 layer thickness plotted as a functions of $(t + \tau)/x$, as calculated using Eq. (3.50). The slope of the data gives constant, *B*, and the vertical intercept means –*A*. The activation energy for the Si wafer oxidation could be calculated from the temperature dependence of the slope. In addition, the activation energy for the interfacial chemical reaction can be obtained from the temperature dependence of *B*/*A*. Equation (3.50) is known as the Deal-Grove equation, which is applied to oxidation, reduction, and corrosion reactions.

3.4 Sintering [12–15]

3.4.1 Features of Sintering

Sintering was defined by Herring as "morphological changes of particle and pore during heating of fine powder and agglomerate at high temperature below its melting point" and is an important process for fabricating ceramics from powder compacts. Sintering is irreversible and is driven by thermodynamically reducing the excessive energies of particle surfaces, grain boundaries, and pore surfaces. During sintering, particle and pore morphologies change by transfers of atoms and ions to reduce the total surface energies. Figure 3.10 shows schematic illustrations of densification and grain growth by sintering. The number of particle surfaces and pores are diminished, leading to densification. Grain growth reduces the excess grain boundary energies, and these phenomena seem to simultaneously occur during sintering.

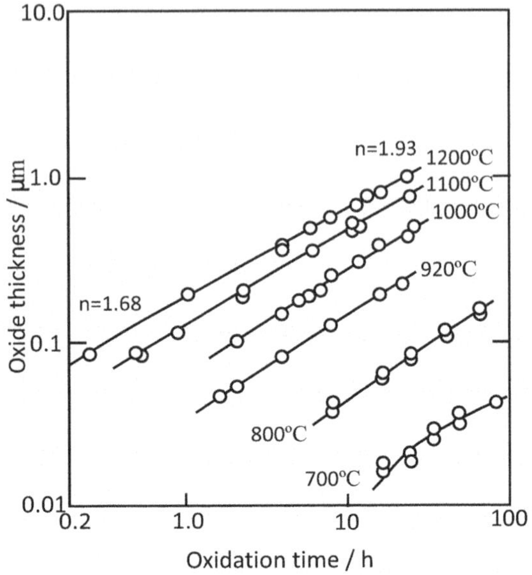

Fig. 3.8 Oxidation of Si wafer under dried oxygen atmosphere at various temperatures. (It is modified from Ref. [11]

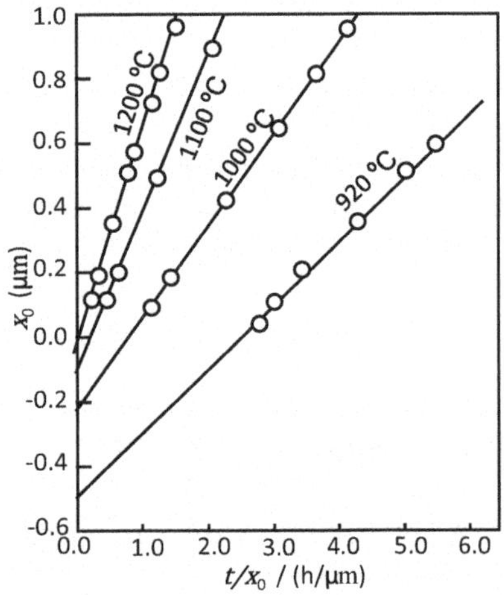

Fig. 3.9 Evaluation of oxidation rate constants of a Si wafer under wet oxygen atmosphere at various temperatures. Slope of line correspond to B, and intercepts correspond to A. (It is modified from Ref. [11]

The densification by sintering causes ~20% linear shrinkage. A shrinkage curve is often measured to examine the sintering process. The linear shrinkage during sintering is defined as the ratio of the shrunken length (L_0-L) to the initial length (L_0), where L is the length of the sample at time t and is measured by thermodilatometry. Sintering can be classified as solid state sintering, liquid phase-assisted sintering, and viscous sintering. In addition, the sintering processes assisted by mechanical pressure, gas pressure, electrical fields, etc. have been developed. Sintering mechanisms and rates also have been extensively studied. Classical sintering theory will be described later.

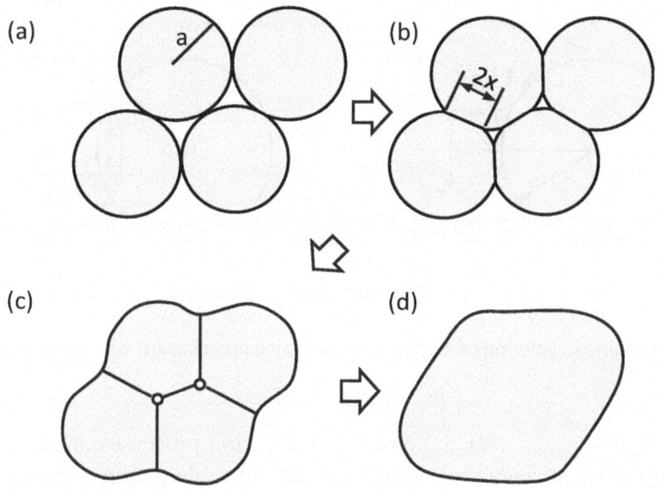

Fig. 3.10 Schematic illustration of sintering. (**a**) Particle packing, (**b**) necking between particles, (**c**) neck growth and shrinkage of pores, and (**d**) grain growth

In practice, sintering is affected by intrinsic and extrinsic factors. Particle packing homogeneity and some additives are examples of intrinsic factors and atmospheric, and external pressures are examples of extrinsic factors that significantly affect the sintering. Although sintering is often studied from the scientific view of interest, it must be studied from the practical and industrial perspectives for application to manufacturing processes.

3.4.2 Sintering Mechanism

Packed powder compacts densify and grains grow during sintering. We will now discuss the mechanisms of atomic- and ionic-scale solid state sintering using the two-particle model as shown in Fig. 3.11. Figure 3.11a shows the model of sintering without any shrinkage, and Fig. 3.11b shows densification with shrinkage. Some migration and diffusion routes of atoms are considered in each model as follows:

A. Vaporization-condensation
B. Surface diffusion
C. Volume (bulk) diffusion or lattice diffusion
D. Grain boundary diffusion
E. Viscous flow and creep flow

The diffusion processes (i.e., routes B, C, and D) are the typical mechanisms by which atoms and ions transfer on surface, along grain boundary, and inside bulk volume. The diffusion coefficients in these routes are $D_{surface}$, $D_{grain\ boundary}$, and D_{volume}, respectively. They depend on the diffusion route in the order of

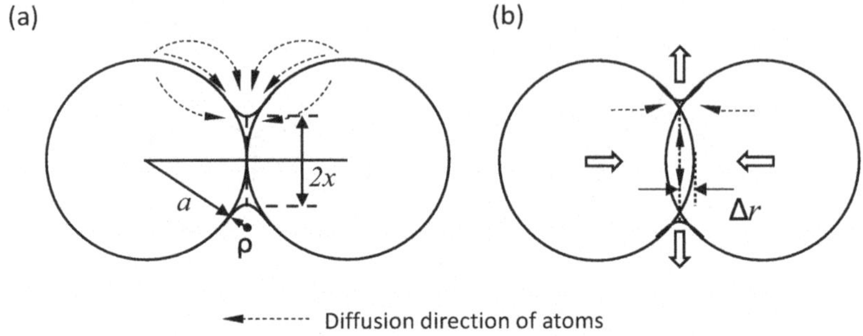

Fig. 3.11 Schematic illustration of sintering between two particles: (**a**) without shrinkage, (**b**) with shrinkage

$D_{surface} > D_{grain\ boundary} > D_{volume}$, because diffusion processes are associated with vacancy diffusion. Surface atoms are basically unstable owing to lack of bonding, and vacancies exist along grain boundary because of dangling bonding. Although atoms and ions can easily transfer through surfaces and grain boundaries at lower temperature, it is difficult to diffuse through lattices at low temperature. To accelerate atomic diffusion, additional energy such as heat is required in order to transfer atoms. At high temperature, lattice vibrations become vigorous, so vacancy concentrations exponentially increase. Therefore, bulk volume diffusion occurs at high temperature and determines densification rates. The lattice diffusion is the volume diffusion from inside defects.

Atomic and ionic diffusion rates are proportional to both vacancy concentrations and concentration gradients. In regions showing high vacancy concentrations, atoms and ions transfer to reduce vacancy concentration gradients. In the model shown in Fig. 3.11, the vacancy concentration around the neck of two particles with concave curvature is higher than that at the particle surface with convex curvature. Although the surface tension induces the compressive stress to inside direction at the particle surface, it induces the tensile stress to outside direction at the neck surface. Therefore, the vacancy concentration is low at the particle surface and high at the neck surface. This vacancy concentration gradients cause the vacancy diffusions from the neck to the particle surface. On the contrary, atoms and ions transfer in the opposite direction, resulting in the neck growth, that is, extension of grain boundary. This is the volume diffusion from the particle surface. However, it does not contribute to the densification, because the distance between two particles does not change although the particle itself shrinks. This is the case in Fig. 3.11a. To cause the densification, the migration of atoms from particle inside to the neck should take place, and especially, the grain boundary plays an important role. There are two ways: one is the volume diffusion through inside lattice from grain boundary, and another is the grain boundary diffusion at the grain boundary. In these processes, the inside atoms are consumed to the neck growth, and two particles come close to each other, leading to the densification with shrinkage. This is the case in Fig. 3.11b. The surface diffusion does not contribute to the densification in the same meaning of volume diffusion from the particle surface (Fig. 3.11a). The surface diffusion proceeds at low temperature and reduces the sintering activity of fine particles. The long-time soaking at low temperature should be avoided.

For vaporization and condensation, the mass transfer depends on the surface curvatures and vapor pressure of the particle. The well-known relationship between curvature and vapor pressure is defined by the Kelvin equation as follows:

$$\ln\frac{P_r}{P_0} = \frac{V_A\gamma}{RT}\left(\frac{1}{r_1} + \frac{1}{r_2}\right), \tag{3.59}$$

where P_0 and P_r are the vapor pressures on the planar and curved surfaces, respectively, V_A is the molar volume, γ is the surface energy, R is the gas constant, and T is the absolute temperature. The curvature on the curved surface is expressed by two curvature radii, r_1 and r_2, on ellipsoid surface. The particle surface has concave curvature with positive curvature radii. At the neck surface, the curvature radius in the circumference direction of the contact area is positive, while the curvature radius in the perpendicular direction is negative and small, which makes a large effect as concave curvature as mentioned in Fig. 3.12. The surface energies of convex curvatures are higher than those of concave surfaces, so atoms tend to vaporize from convex surfaces to concave ones in order to reduce surface energies. Although this mechanism contributes to particle deformation, it does not contribute to the densification similarly to the surface diffusion as seen in Fig. 3.11a. It makes a problem for sintering of volatile materials.

The viscous/creep flow mechanism appears in the sintering of glass particles. Glass phase does not have crystal lattice and forms an amorphous network structure. The constituent atoms do not migrate by independent diffusion mechanism and, rather, migrate by a local collective movement. The driving force of the viscous sintering is the difference in stress between adjacent particles. It induces the viscous flow, leading to the densification.

3.4.3 Sintering Shrinkage

3.4.3.1 Sintering Steps

Sintering can be classified as initial-, mid-, and final-stage densifications for determining shrinkage ratios. During initial-stage sintering, relative sample densities are 60–65% (or more) higher than those of powder compacts, and linear shrinkages are <5%. The contact areas, called "necks," between adjacent particles are formed. The contact area expands with progress of sintering. It is called "neck growth," and the distance between particles shrinks by diffusion mechanisms involved with grain boundary. During middle stage of sintering, the neck growth is promoted, the linear shrinkage is enlarged to 10–15%, and the relative density reaches 90–95%. Along with the densification, fine pores diminish and/or incorporate into large pores. Some

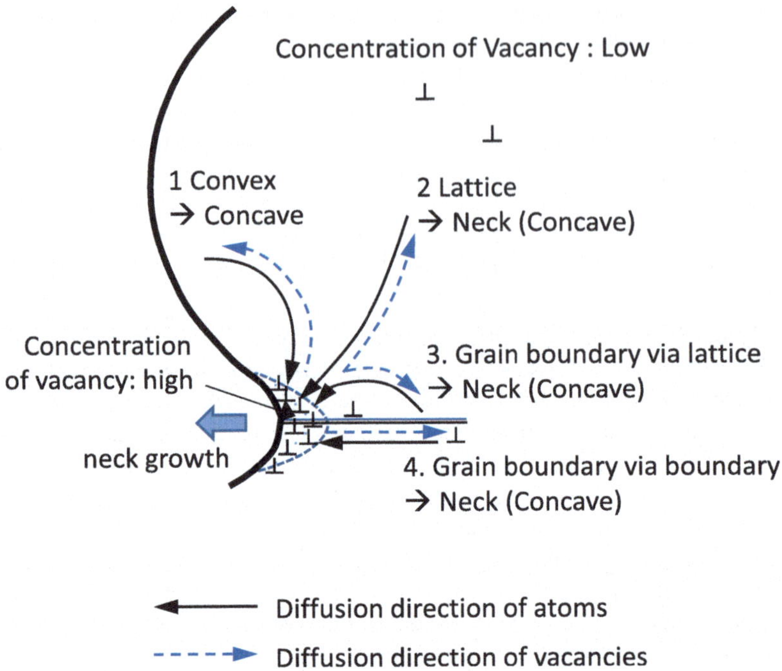

Fig. 3.12 Schematic illustration of diffusion of atoms and vacancies on neck growth

pores connect to each other, forming pore network. When the relative density is above 90%, the grain growth occurs, and the residual pores change to isolated closed pores. Actually, since particle sizes are distributed and local particle densities slightly fluctuate in powder compacts, sintering is very complicated because many sintering mechanisms are simultaneously happening. During final-stage sintering, relative densities reach > 98%, so grains mainly grow. In the next section, linear shrinkage is introduced for each sintering mechanism.

3.4.3.2 Analysis of Sintering Kinetics [2, 13–15]

Numerous researchers have proposed sintering mechanism. They assumed the two-particle model and considered the geometry of connected particles, the flux of atoms, the counter flow of vacancies, etc. Neck growth during initial stage of sintering, which is part of classical sintering theory, has been discussed. Herein, the sintering rate during initial stage is introduced. First, the kinetics of volume diffusion process are discussed, and then the other kinetics are briefly introduced.

Here, J is the atomic flux into the neck, c is the number of atoms per unit volume, and v is the atomic transfer rate [13].

$$J = cv. \quad (3.60)$$

Further, v is expressed by the driving force, F, and the mobility, B, so Eq. (3.60) becomes

$$J = cFB, \quad (3.61)$$

and since F is equal to the chemical potential gradient of vacancy, $d\mu/dx$, Eq. (3.61) can be written as

$$J = cB\frac{d\mu}{dx}. \quad (3.62)$$

Since the chemical potential is given by $\mu = \mu_0 + RT \ln P$,

$$J = cBN_A\Omega\frac{dP}{dx}, \quad (3.63)$$

where P is the pressure induced by surface tension, N_A is the Avogadro number, and Ω is the atomic volume. Here, the absolute temperature, T, is constant, so Fick's first law is given by

$$J = -D\frac{dc}{dx}, \quad (3.64)$$

where D is the diffusion coefficient. By comparing Eqs. (3.63) and (3.64),

$$B = \frac{D}{N_A kT}, \quad (3.65)$$

where k is the Boltzmann constant. By substituting Eq. (3.65) into Eq. (3.63) and considering the atomic volume Ω is equal to $1/c$, we obtain

$$J = -\frac{D}{kT}\frac{dP}{dx}. \quad (3.66)$$

This equation indicates that the pressure difference due to surface tension induces the diffusion of atoms.

For lattice diffusion from grain boundary, the volume of matter transported into the neck per unit time is given by

$$\frac{dV}{dt} = JA\Omega, \quad (3.67)$$

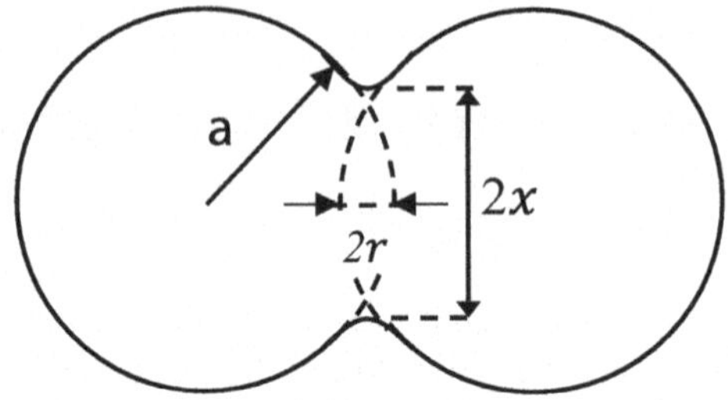

Fig. 3.13 Geometrical illustration of connected particles

where J is the flux of the transported matter, as in Eq. (3.66), and corresponds to the volume diffusion coefficient, D_v. A is the cross-sectional area over which the diffusion occurs. The geometry of the two-particle model shown in Fig. 3.13 is as follows:

$$\text{Radius of neck concave curvature} \quad r = \frac{x^2}{4a}, \tag{3.68}$$

$$\text{The cross} - \text{sectional area of neck} \quad A = 2\pi xr, \tag{3.69}$$

$$\text{The volume of neck} \quad V = \pi x^2 2r. \tag{3.70}$$

By combining Eqs. (3.67), (3.68), (3.69), and (3.70), we obtain

$$\frac{dV}{dt} = \frac{D_v}{kT} 2\pi xr\Omega \frac{dP}{dx}. \tag{3.71}$$

The pressure on curved surface is expressed by the Laplace's equation ($P = \gamma_s/r$), where γ_s is the surface energy and r is the curvature radii, so the change of neck radius is

$$\left(\frac{x}{a}\right)^4 = \frac{16 D_v \gamma_S \Omega}{a^3 kT} t. \tag{3.72}$$

The change of shrinkage $\Delta L/L_0$ is

$$\frac{\Delta L}{L_0} = \frac{r}{a} = \frac{x^2}{4a^2} = \left(\frac{D_v \gamma_S \Omega}{a^3 kT}\right)^{1/2} t^{1/2}. \tag{3.73}$$

When atoms are transferred through grain boundaries to neck growth regions, i.e., by grain boundary diffusion routes, the sintering kinetics are as follows:

$$\frac{dV}{dt} = 2\pi x \delta_{gb} \Omega \frac{D_{gb}}{kT} \frac{\gamma_S}{a^4}. \tag{3.74}$$

By combining Eqs. (3.70) and (3.74) and integrating the resulting equation, the change of neck radius is obtained as

$$\left(\frac{x}{a}\right)^6 = \frac{96\delta_{gb}D_{gb}\gamma_S\Omega}{kTa^4}t, \tag{3.75}$$

and the linear shrinkage, $\Delta L/L_0$, is expressed as follows:

$$\frac{\Delta L}{L_0} = \left(\frac{3\delta_{gb}D_{gb}\gamma_S\Omega}{4kTa^4}\right)^{1/3} t^{1/3}. \tag{3.76}$$

Equations (3.73) and (3.76) show that the sintering shrinkage is proportional to $t^{1/2}$ or $t^{1/3}$ as a function of time. The measured shrinkage is plotted against sintering time in log-log scale, that is, in the formula of log ($\Delta L/L_0$) = log k + n log t where k is defined as rate constant. From the slope, n, we can determine whether volume diffusion governs the sintering or grain boundary diffusion. Furthermore, the rate constant, k, is determined from the intercept, and the activation energy can be estimated from the temperature dependence. From the term of rate constant in Eqs. (3.73) and (3.76), the sintering rate is considerably affected by the particle size. Figure 3.14 demonstrates the effect of particle size on the sintering of silicon carbide [16]. The densification is enhanced with a decrease in particle size and the relative density reaches > 92% with fine particles of < 0.02 μm by pressureless sintering. For sintering kinetics through other diffusion mechanisms during initial stage of sintering, the rates of neck growth are expressed in Table 3.1.

As sintering proceeds, the pore channels between particles are disconnected and isolated pores formed. Coble proposed two geometrically simple models for analyzing pore volume change rate during middle and final stage of sintering. The geometrical model of middle stage of sintering is bcc-packed tetrakaidecaheral grains with cylinder-shaped pores along grain edges as shown in Fig. 3.15a. Pore volume change rates (P) during middle sintering stage by lattice diffusion and grain boundary diffusion are expressed, respectively, in Eqs. (3.77) and (3.78) as follows:

$$P = \frac{10D_v\gamma_S\Omega}{l^3kT}\left(t_f - t\right), \tag{3.77}$$

$$P = \left(\frac{2\delta_{gb}D_{gb}\gamma_S\Omega}{l^4kT}\right)^{2/3} t^{2/3}, \tag{3.78}$$

where l is length of grain edge. For final sintering stage, pore volume change rate is induced based on the bcc-packed tetrakaidecaheral grains with spherical-shaped

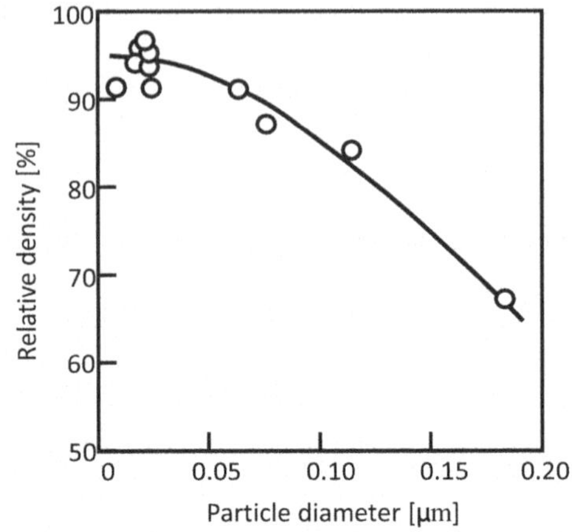

Fig. 3.14 Dependence of relative density on particle size of silicon carbide ceramics. (Sintering temperature, 2050 °C; duration time, 30 min; Ar atmosphere; sintering additives, boron (1 wt%)

pores at their corners as shown in Fig. 3.15b. The pore volume change rate is expressed as follows:

$$P = \frac{6\pi D_v \gamma_S \Omega}{\sqrt{2} l^3 kT} \left(t_f - t \right). \tag{3.79}$$

3.4.4 Kinetics of Grain Growth [17]

The control of grain size is very important. The grain sizes produced during final-stage sintering govern the functional properties (i.e., the electrical, magnetic, optical, and mechanical properties) of polycrystalline ceramics.

The grain growth occurs and fine pores diminish simultaneously during middle or final stage of sintering. Figure 3.16 illustrates the two-dimensional model of microstructure and grain growth in polycrystalline ceramics, where many grains with various sizes exist in the ceramics. The thermodynamically stable grain shape is hexagon surrounded by six adjacent grains, because the bond angle of grain boundaries at the triple point is 120° owing to the balance of grain boundary tension. However, most of grains are irregular in shape, and the grain boundary is curved depending on the number of surrounding grains. Small grains are surrounded by small number of large grains, and the grain boundaries are concaved. In contrast, large grains are surrounded by many small grains, and the grain boundaries are convex. The compressive stress is induced from concave side to convex side owing to the grain boundary tension, resulting in the migration of grain boundary from large grain to small grain. Large grains usually grew by incorporating fine grains. This is the grain growth. This phenomenon is similar to "Ostwald ripening."

Table 3.1 Kinetic equations for various mechanisms of initial-stage sintering

Mechanism	Neck growth rate	Linear sintering shrinkage rate	References
Surface diffusion	$\left(\frac{x}{a}\right)^7 = \left(\frac{56D_s\gamma_s\Omega}{a^4}\right)t$	$\frac{\Delta L}{L_0} = 0$	Kuczynski [18]
Lattice diffusion from grain boundary to neck	$\left(\frac{x}{a}\right)^4 = \left(\frac{16D_v\gamma_s\Omega}{a^3kT}\right)t$	$\frac{\Delta L}{L_0} = \left(\frac{D_v\gamma_s\Omega}{a^3kT}\right)^{1/2}t^{1/2}$	
Grain boundary diffusion from grain boundary to neck	$\left(\frac{x}{a}\right)^6 = \left(\frac{96\delta_{gb}D_{gb}\gamma_s\Omega}{a^4kT}\right)t$	$\frac{\Delta L}{L_0} = \left(\frac{3\delta_{gb}D_{gb}\gamma_s\Omega}{4a^4kT}\right)^{1/3}t^{1/3}$	Coble [17]
Lattice diffusion from particle surface to neck	$\left(\frac{x}{a}\right)^5 = \left(\frac{40D_v\gamma_s\Omega}{a^3kT}\right)t$ $\left(\frac{x}{a}\right)^5 = \left(\frac{80D_v\gamma_s\Omega}{a^3kT}\right)t$	$\frac{\Delta L}{L_0} = \left(\frac{20D_v\gamma_s\Omega}{\sqrt{2}a^3kT}\right)^{2/5}t^{2/5}$	Kingery and Berg [15]
Evaporation-Condensation	$\frac{x^3}{a} = Kt$	$\frac{\Delta L}{L_0} = 0$	Kingery and Berg [15]
Viscous flow	$\left(\frac{x}{a}\right)^2 = \frac{3\gamma_s}{2a\eta}t$	$\frac{\Delta L}{L_0} = \frac{3\gamma_s}{8\eta a}t$	Frenkel and Eshelby [19, 20]

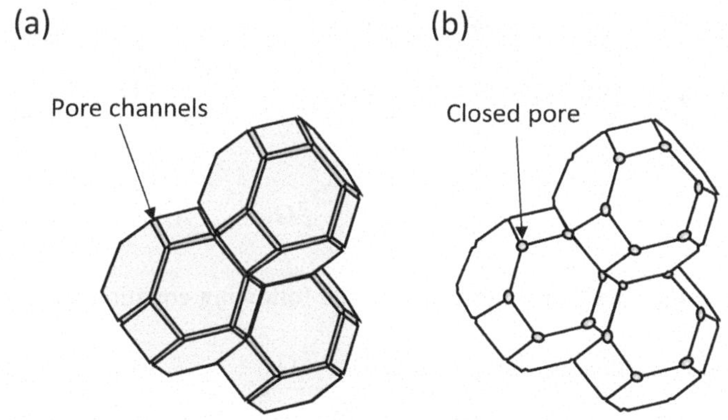

Fig. 3.15 Sintering geometrical models for (a) middle stage and (b) final stage

During the grain growth, atoms transfer across the grain boundary from small grain to large grain. The grain growth rate (*dG/dt*) is proportional to the mobility of grain boundary, which is expressed as a function of the atomic flux according to the pressure gradient as follows:

$$\frac{dG}{dt} = \alpha J\Omega = \alpha\frac{D_b^{\perp}}{kT}(\Delta P)\Omega, \tag{3.80}$$

where α is the constant, J is the atomic flux, Ω is the atomic volume, $D_b^{⍰}$ is the coefficient for diffusion across the grain boundary, k is the Boltzmann constant, and T is the absolute temperature. ΔP is the pressure gradient of grain boundary and is expressed as a function of the grain boundary energy γ_b and the average radius of curvature, $\bar{R}_0$, and Eq. (3.80) is rewritten as follows:

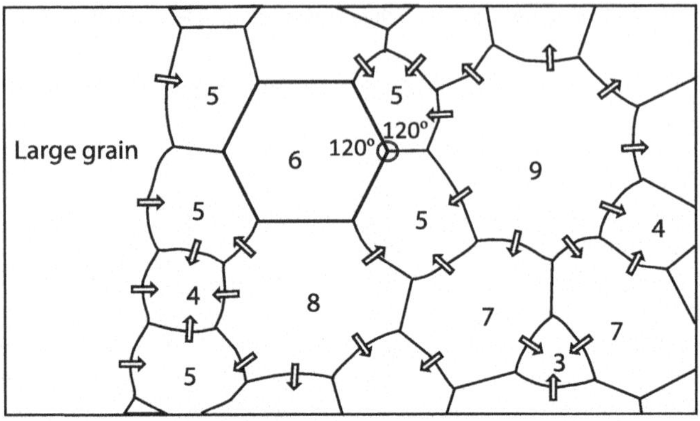

Fig. 3.16 Microstructure and migration of grain boundary during sintering. The arrows mean the direction of grain boundary migration

$$\frac{d\bar{G}}{dt} = \alpha \frac{D_b{}^{\perp}}{kT} \frac{2\gamma_b}{\bar{R}_0} \Omega. \tag{3.81}$$

Since $\bar{R}_0$ is proportional to the average grain size $\bar{G}$, Eq. (3.81) can be rewritten as follows:

$$\frac{d\bar{G}}{dt} = \beta \frac{D_b{}^{\perp}}{kT} \frac{2\gamma_b}{\bar{G}} \Omega. \tag{3.82}$$

By integrating Eq. (3.82) from time t_0 to t, the following equation is obtained:

$$\bar{G}_t{}^2 - \bar{G}_0{}^2 = \frac{4\beta D_b{}^{\perp}\gamma_b \Omega}{kT} \mathrm{t} = Kt. \tag{3.83}$$

The grain growth rate is proportional to the square root of sintering time. When log $\bar{G}_t$ is plotted as a function of log$t$ on the assumption of $G \gg G_0$, the straight line is obtained with a slope of 1/2. The activation energy of the grain growth is obtained from the temperature dependence of K. For the grain growth in ceramics, the rate equation is often expressed by the following equation considering the other factors such as pores or inclusions segregated at the grain boundary:

$$\overline{G_t}^m - \overline{G_0}^m = K't. \tag{3.84}$$

The order index, *m,* is given ordinarily as 3.

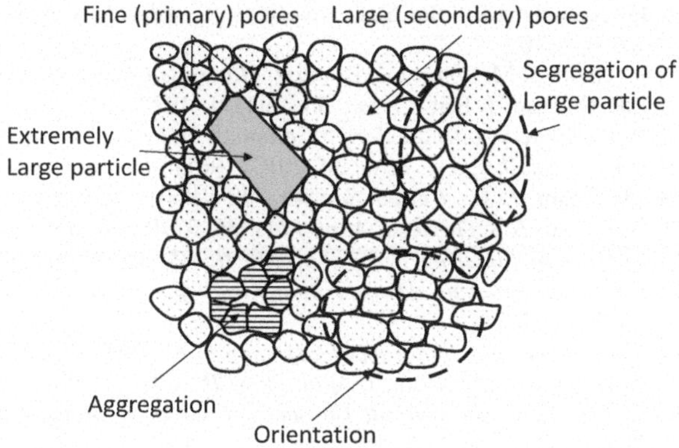

Fig. 3.17 Schematic illustration of various packing structures in practical powder compact

3.4.5 Microstructure

In practical ceramic processing, the powder compact consists of various levels of particle-packing structures as seen in Fig. 3.17, such as large and small agglomerates, extremely large particle, oriented agglomerate, and so on. For pore structures, there are two types: fine pores among primary particles and large pores surrounded by secondary agglomerates. Fine pores diminish during the shrinkage by sintering, while large pores are difficult to eliminate, remain in the sintered body, and cause pore growth by incorporation of fine pores. As mentioned above, the classical sintering theory is usually used to analyze the sintering shrinkage in two or some particle model. To understand the sintering in practical packing structure, these inhomogeneous structures must be considered in the model. Besides, the constrained sintering must be considered in various applications. Sintering of particles layered thinly on the rigid substrate is conducted for electrical devices or surface coatings. The particles sinter under constrained force from the substrate. If the force exceeds limitation, the delamination or cracking happens during the sintering. Thus, recent studies on sintering are expanding to understand and control the microstructure toward various applications. Main topics are (1) inhomogeneity and its influence on microstructure and shrinkage; (2) constrained sintering, effects of substrate and/or rigid structures; (3) morphological stabilities of mixed phases and continuous pores; (4) effect of sintering additives; (5) effect of process variables; and (6) viscous sintering, etc.

References

1. P.D. Garn, T.S. Habash, J. Phys. Chem. **85**, 229 (1979)
2. W.D. Kingery, *Introduction to Ceramics,* 2nd edn. (Wiley, New York, 1975), p. 321
3. P. Shewmon, *Diffusion in Solids*, 2nd edn. (Wiley, New York, 1989), p. 9
4. F.D. Richardson, J.H.E. Jeffes, *J. Iron Steel Inst.* **160**, 261 (1948)

5. H. Schmalzried, Solid State Reactions, Translated by A. D. Petlon, (Academic, Cambridge, 1974), p. 109
6. A. Ghorbel, M. Fourati, J. Bouaziz, *Materials Chemistry and Physics* **112**, 876 (2008)
7. S.F. Hullbert, J. Brit, *Ceram. Soc.* **6**(1), 1 (1972)
8. W.D. Kingery, *Introduction to Ceramics*, 2nd edn. (Wiley, New York, 1975), p. 413
9. R.E. Carter, *J. Chem. Phys.* **34**, 2010 (1961)
10. J.H. Sharp, *J. Am. Ceram. Soc.* **49**, 379 (1966)
11. B.E. Deal, A.S. Grove, *J. Appl. Phys.* **36**, 3770 (1965)
12. R. German, *Sintering Theory and Practice* (Wiley, New York, 1996), p. 67
13. S.J. Kang, *Sintering, Densification, Grain Growth and Microstructure* (Elsevior, Amsterdam, 2005), p. 37
14. M.N. Rahamann, *Sintering of Ceramics* (CRC Press, Boca Raton, 2008), p. 45
15. W.D. Kingery, M. Berg, *J. Appl. Phys.* **26**, 1205 (1955)
16. J. Hojo, in *Silicon Carbide Ceramics -1,* ed. by S. Somiya, Y. Inomata (Elsevier, Amsterdam, 1991), p. 163
17. R.L. Coble, *J. Am. Ceram. Soc.* **41**, 55 (1958)
18. G.C. Kuczynski, Self-diffusion in sintering of metallic particles. JOM **1**(2), 169–178 (1949)
19. J. Frenkel, J. Phys. **4**, 385 (1945)
20. J.D. Eshelby, Transaction AIME **185**, 806 (1949)

Chapter 4
Powder and Thin Film Synthesis

Toshinobu Yogo

Abstract Ceramic powders are synthesized using chemical solution methods, such as organic acid salt, precipitant generation, alkaline hydrolysis, and alkoxide hydrolysis. The powder synthesis by precipitation and hydrolysis and the synthesis of monodispersed particles are described. The formation of complex alkoxide of lithium niobate (LN) precursor is confirmed by ^{93}Nb nuclear magnetic resonance spectroscopy. Stoichiometric LN films are synthesized on various substrates via chemical solution route. The synthesis of the designed precursor in solution is a key for the low-temperature crystallization of high-quality LN films with preferred orientation. Several functional ceramic thin films of multicomponent oxides are also synthesized using chemical solution process.

Keywords Chemical solution process · Monodispersed particle · Complex alkoxide · Epitaxial film

4.1 Fabrication Processes for Ceramic Powders and Films

Ceramic powders are mainly synthesized via three methods: solid phase, liquid phase, and gas phase reaction. In the solid phase method, a raw material oxide is calcined at temperatures of 1000 °C or above, and a highly crystalline product is obtained. In the liquid phase method, raw material components dissolved into a liquid solution are used as starting material, and the synthesized ceramic powder component is homogenous and has an active particulate surface and controllable powder characteristics. In the gas phase method, since the starting compounds are reacted in an inert gas, a high-purity product is more easily obtained. The gas phase also enables synthesis of fine powders of nitrides, carbides, etc. which are difficult to synthesize using liquid phase reactions.

T. Yogo (✉)
Institute of Materials and Systems for Sustainability, Nagoya University, Nagoya, Japan
e-mail: yogo@chembio.nagoya-u.ac.jp

J. Hojo (ed.), *Materials Chemistry of Ceramics*,
https://doi.org/10.1007/978-981-13-9935-0_4

Two major methods are used for ceramic thin film synthesis: physical methods and chemical methods. Physical methods include heating thin film raw materials in a vacuum apparatus or vaporizing the thin film raw material via sputtering, laser ablation, etc., to deposit the vapor particulate onto a substrate to make the film. Chemical methods consist of vapor phase growth methods and chemical solution methods. In vapor phase growth, thin film raw materials are reacted with halides or hydrides in the gas phase, and the thin film compound is vapor-deposited onto a substrate. In the chemical solution method, a film is formed on a substrate using different methods of coating of the raw material solution and crystallized to form an oxide thin film.

This chapter describes the synthesis methods of ceramic powders and thin films with the liquid phase method using chemical reactions in solution.

4.2 Powder Synthesis with the Chemical Solution Method

4.2.1 Characteristics of Synthesis with the Chemical Solution Method

Crystallization of a product is often used to realize desired physical characteristics in ceramic materials. To do so, the starting compounds must undergo heat treatment at high temperatures of 1000 °C or above. In the ordinary solid phase method, a mechanical mixture of raw material oxide powder and other materials is calcined; however, the component elements are not mixed homogeneously. Even after high-temperature processing, segregations and second-phase remnants can be present.

In contrast, by controlling solution reactions in the chemical solution method, the component elements can be mixed in the preheat treatment stage to homogenize them at the molecular level, thereby resolving the above-described problem points. The chemical solution method is characterized as follows: crystal growth is possible at low temperatures, component control is easy, homogeneous additions of trace elements are possible, and there is high homogeneity of product. Another major characteristic of the solution method is that, in addition to powders, porous materials, films, fibers, and a variety of other material configurations, can also be synthesized. The preparation of such films and fibers from solid phase reactions using oxides is difficult.

4.2.2 Powder Synthesis via Precipitation

Figure 4.1 shows the classifications of ceramics precipitated from solution and formed by desolventizing. One ceramic powder synthesis reaction from solution is the precipitation reaction, where coprecipitation, hydrolysis, etc. are used to obtain a multicomponent hydroxide powder. Spray-drying and freeze-drying are means of removing the solvent from a solution containing metal ions. Another liquid phase

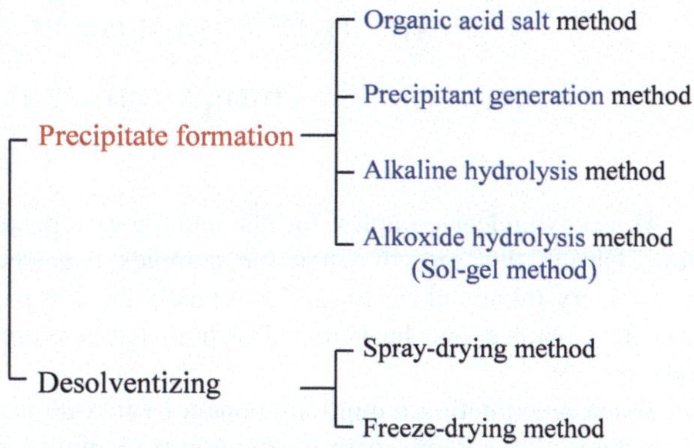

Fig. 4.1 Classifications of ceramics precipitated from solution and formed by desolventizing

method is the hydrothermal technique for synthesizing a single crystal such as quartz under pressure; this technique is also used to synthesize nanocrystalline powders. This section describes precipitation reactions under normal pressure.

To precipitate a powder from solution, in some cases the precipitant is added from outside, and in others the precipitate is formed within the solution via chemical reactions. For metal-organic acid salt method, the precipitants, such as oxalate and citrate, are added externally. When sodium citrate is added to an aqueous solution with a dissolved mineral salt, a metal citrate salt is precipitated [1]. For example, when aqueous oxalic acid solution is reacted with an aqueous solution containing barium chloride and titanium oxychloride solution such that the Ba/Ti ratio is 1.0, a white $BaTiO(C_2O_4)_2 \cdot 4H_2O$ precipitate is synthesized. Stoichiometric barium titanate ($BaTiO_3$) powder is prepared by heating this precipitate in air and removing its organic constituent. This powder has superior sinterability and is used as a high-purity, dielectric base powder.

The precipitant generation method involves making the precipitant within a solution and causing a homogeneous precipitation reaction. One such precipitant is ammonia made by hydrolysis after heating an aqueous urea solution above 70 °C [2]. Precipitant generation can be controlled via selection of urea concentration, temperature, pH, etc.

$$(NH_2)_2CO + H_2O \rightarrow 2NH_3 + CO_2 \tag{4.1}$$

Other precipitants are diethyl sulfate, thiourea, etc., used for precipitant generation of sulfate and sulfide, respectively.

In the alkaline hydrolysis method, alkali is added to an acidic metal salt solution, which then undergoes hydrolysis. A metal salt such as nitrate has high solubility in water, and a homogenous solution can be prepared and includes the target component elements. Making this solution alkaline enables precipitation of a metal hydroxide or oxide. Anions and H_2O coordinate with metal ions in the solution to form complex ions. These complex ions form a polynuclear complex via hydrolysis and condensation reactions. For example, Al^{3+} forms a hydrated ion in aqueous solution and forms a polynuclear complex such as $[AlO_4Al_{12}(OH)_{24}(H_2O)_{12}]^{7+}$ [3].

$$\mathrm{Al(H_2O)_6}^{3+} \rightarrow \left[\mathrm{Al(H_2O)_5OH}\right]^{2+} + \mathrm{H^+} \tag{4.2}$$

$$\begin{aligned} 2\mathrm{Al(H_2O)_6}^{3+} &\rightarrow \left[\mathrm{(H_2O)_5Al(OH)_2Al(H_2O)_5}\right]^{4+} + 2\mathrm{H^+} \\ &\rightarrow \ldots\ldots \rightarrow \left[\mathrm{AlO_4Al_{12}(OH)_{24}(H_2O)_{12}}\right]^{7+} \end{aligned} \tag{4.3}$$

The polynuclear complex further undergoes repeated association and dissociation. During this growth when the complex exceeds a certain size (the critical radius), crystal nuclei are formed and precipitated as particles. Precipitated particles can be separated and heat-treated at high temperatures to obtain the metal-oxide powder.

When precipitating a multicomponent hydroxide from a solution including multiple types of metal ions, careful attention is required for differences in the solubility products of respective metal hydroxides. When there are large differences in the solubilities of the metal ions due to pH, one must be aware that the precipitate composition may not match the solution composition.

The alkoxide hydrolysis method is also called the sol-gel method. A metal alkoxide is a compound made of M—O—R bonds (M, metal element; R, alkyl group) and is expressed by $M(OR)_n$. The M—O bond is a bond between an electropositive metal and a negative oxygen and is readily susceptible to hydrolysis, yielding an oxide as follows [4]:

Hydrolysis reaction

$$\mathrm{M(OR)_n + H_2O \rightarrow M(OR)_{n-1}(OH) + ROH} \tag{4.4}$$

$$\mathrm{M(OR)_{n-1}(OH) + (n-1)H_2O \rightarrow M(OH)_n + (n-1)ROH} \tag{4.5}$$

Polycondensation reaction

$$\mathrm{mM(OH)_n \rightarrow mMO_{n/2} + nm/2H_2O} \tag{4.6}$$

The formation of a complex alkoxide by the reaction of two or more metal alkoxides is also known. Synthesis of an appropriate complex alkoxide and its use as a starting material enable attainment of a complex oxide as shown in the following formula:

$$\mathrm{M(OR)_n + M'(OR)_m \rightarrow MM'(OR)_x}$$

Hydrolysis and polycondensation

$$\rightarrow \ldots\ldots \rightarrow \mathrm{MM'O_y} \tag{4.7}$$

The alkoxide hydrolysis product easily forms a network structure, and after passing through a sol phase, it changes to a gel. Using these structural changes, inorganic materials having a variety of shapes can be synthesized. Three-dimensional structures of products can be controlled by adjusting the alkoxide hydrolysis conditions. Hydrolysis of alkoxysilane (silicon alkoxide, $Si(OR)_4$) is known to progress under acidic conditions via electrophilic substitution reactions and under basic conditions by nucleophilic substitution reactions [5]. Under acidic conditions, H_3O^+ electrophilically attacks the oxygen of the Si—OR bond alkoxy groups, and with the elimination of ROH, Si—OH bonds are formed. Si—OH bonds condense with other Si—OH bonds, and a chain polymer is formed. Under basic conditions, however, OH^- ions nucleophilically attack Si, and OR^- groups are eliminated. Due to steric hindrance of the substituent groups around the central Si, "bulkier" R groups show a more rapid decrease in reactivity. Generally, under acidic conditions, a network structure develops, and it is easy for a gel to form, while under basic conditions hydrated particles are produced. In this way, the hydrolysis reaction rate and resultant product of alkoxide depend on a variety of factors including acidity, basicity, as well as substituent type, solvent, reaction temperature, water quantity used for hydrolysis, concentrations, etc. Appropriate selection of these conditions enables control of the product's three-dimensional structure, as well as adjustment of sol viscosity. Spinnable gel fibers can be obtained from a precursor sol whose viscosity has been controlled, enabling preparation of ceramic fibers.

4.2.3 Synthesis of Monodispersed Spherical Particles

Important particle properties are shape, particle size, and particle size distribution. These properties impact the electrical and optical properties of the particles themselves. A well-known example is that the coercivity of a magnetic particle depends on its particle size (particle diameter). When the particle size decreases and its size changes from a multi-magnetic domain structure to a single domain structure, a large coercive force is observed [6]. Another example of size effect is that the optical absorption edge of a semiconductor particle is dependent on particle size, and at nanosizes, the short wavelength shift also depends on the particle size [7]. In the preparation of a ceramic sintered body, to ensure that it has the desired properties, it is necessary to control the particle size and size distribution of the raw material particles. When particles of submicron size have a narrow size distribution, the sintering temperature can be lowered, and a dense sintered body with controlled microstructure can be prepared. Thus, particles with a controlled particle shape, size, and size distribution are required to meet a desired goal. In a solution reaction, monodispersed spherical particles with a sharp size distribution can be prepared by controlling nucleation and growth.

Figure 4.2 shows the relationship between time and solute concentration when synthesizing monodispersed spherical particles [8]. Here, "solute" means the precursor materials such as polynuclear complexes and polycondensates dissolved in the solution prior to precipitation. As the solute concentration changes, crystal nuclei form and particle growth occurs. In the initial growth process, t_n is the time from when the solute concentration rises and exceeds the degree of saturation C_s, until it

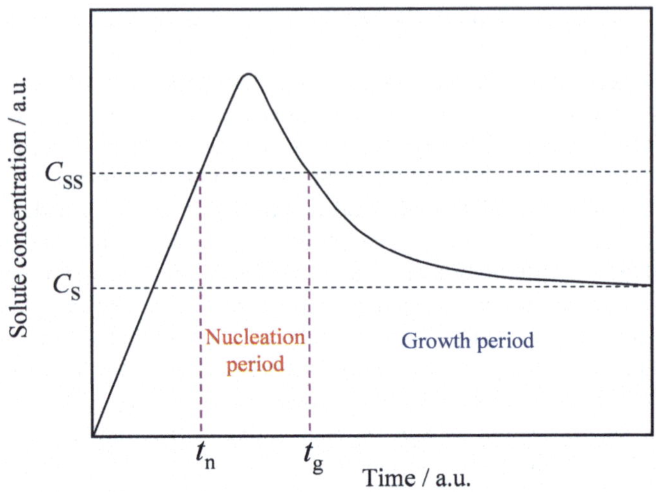

Fig. 4.2 Relationship between time and solute concentration for the synthesis of monodispersed spherical particle. (Reprinted with permission from Ref. [8]. Copyright 1950 American Chemical Society)

reaches the supersaturated concentration C_{ss}. From this t_n to the t_g stage, self-nucleation occurs, and during the period until the concentration again declines to C_s, the dispersion of chemical species to crystal nuclei occurs leading to crystal growth. Nuclei generated in the first period continue to exist, and their particle diameters (sizes) become bigger than the nuclei generated later. Thus, in the brief time period from t_n to t_g, if large numbers of nuclei are explosively generated to promote uniform growth, monodispersed spherical particles can be produced. In the growth period, small-sized (small-diameter) particles are redissolved and disappear, and the solute component precipitates as large-sized (large-diameter) particles. After passing through these processes, the maximum of the particle size distribution curve changes from small particle diameters to large particle diameters, a process known as "Ostwald ripening."

To speed up the nucleation rate so as to complete nucleation at an early time point, the solute concentration must be raised to speed up particle formation reactions. To slow down the rate of particle growth, one must select conditions such as the supersaturation degree and temperature. The generated nuclei are nanoparticles that agglomerate easily. When particle agglomeration occurs in the early stage and secondary particles are formed, these become polydispersed particles. Since it is extremely difficult to dissociate such agglomerated secondary particles for redispersion, a dispersant is added to the solution beforehand, and this effectively prevents agglomeration. Polyelectrolytes, such as sodium polymethacrylate and ammonium polyacrylate, are used as dispersants. A dispersant is adsorbed at the particle surface, the surface charge is controlled, and—due to steric repulsion of polymer chains and other effects—the dispersion state of nanocrystalline particles is stabilized.

Stöber et al. [9] investigated the hydrolysis of silicon alkoxide and found that controlling the growth of $Si(OH)_4$ produced in the intermediate stage enabled the synthesis of monodispersed spherical silica particles of diameter 0.05 μm–2.0 μm. Small-sized particles were obtained from methanol, and large-sized particles were obtained from butanol. In a comparison of silicon substituents, when $Si(OCH_3)_4$ was used, the reaction time was within 1 min, and silica particles with a diameter of 0.2 μm or less were obtained. With $Si(OC_2H_5)_4$ and a reaction time of 24 h, silica particles of approximately 2 μm were obtained. Monodispersed spherical particles of a variety of functional ceramics including silica, TiO_2, ZrO_2, etc. can be synthesized [10].

4.3 Thin Film Synthesis by the Chemical Solution Method

4.3.1 Thin Film Synthesis

Utilizing the characteristics of the solution method, numerous kinds of functional ceramic thin films are currently synthesized. Examples of electronic functional thin film types currently synthesized are ferroelectrics ($BaTiO_3$, $Pb(Ti,Zr)O_3$) [11], transparent conductors (Sn-doped In_2O_3) [12], superionic conductors ($Na_2O{\bullet}11Al_2O_3$) [13], superconductors (YBa_2CuO_7) [14], and others. Optical functional films synthesized by the solution method include light absorption and colored films (FeO, NiO [15], CeO_2-TiO_2 [16]), reflection coatings (PbO-TiO_2 [17], TiO_2-SiO_2 [18]), second-order nonlinear optical films ($KTiOPO_4$ [19], β-BaB_2O_4 [20], $Ba_2NaNb_5O_{15}$ [21]), etc. As a representative example, the synthesis of a stoichiometric epitaxial $LiNbO_3$ film is described below.

4.3.2 Synthesis of Stoichiometric Epitaxial Lithium Niobate Thin Film

4.3.2.1 Lithium Niobate

The structure of lithium niobate ($LiNbO_3$) has hexagonal closest packing of oxygen ions, and a lithium ion and a niobium ion occupy two-thirds of the octahedral sites. This crystal structure is called an "ilmenite-type structure" and belongs to the trigonal system. Therefore, at room temperature, the lithium and niobium ions are at locations slightly askew of the center of the tetrahedral position, and spontaneous polarization occurs; thus, at room temperature, $LiNbO_3$ is a ferroelectric. Due to its superior piezoelectric and optical properties, $LiNbO_3$ is a functional inorganic material that has been researched for many years for its optical material applications, for example, as a surface acoustic wave (SAW) element, etc.

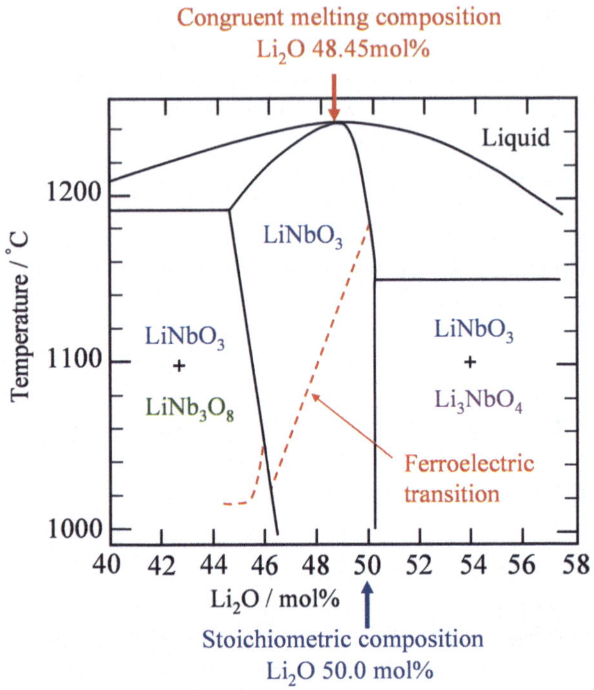

Fig. 4.3 Equilibrium phase diagram of $LiNbO_3$ consisting of lithium oxide (Li_2O) and niobium oxide (Nb_2O_5). (Reprinted with permission from Ref. [22]. Copyright 1971 American Institute of Physics)

Figure 4.3 shows an equilibrium phase diagram of $LiNbO_3$, comprising lithium oxide (Li_2O) and niobium oxide (Nb_2O_5), with magnification of the vicinity of 50 mol% Li_2O [22]. While 50 mol% Li_2O is equivalent to $LiNbO_3$, there exists a solid solution region of width 2%–3% in the vicinity of this stoichiometric composition. The apex of the dome seen at 1250 °C in this phase diagram is called the "congruent melting composition." Detailed research has shown that this congruent melting composition is 48.45 mol% Li_2O. In other words, at 1253 °C, $LiNbO_3$ does not decompose—rather, it is dissolved as is and becomes $LiNbO_3$ melt; that composition, however, is not the 50.0 mol% stoichiometric composition.

The $LiNbO_3$ single crystal is grown using the Czochralski method, by raising a $LiNbO_3$ seed crystal from $LiNbO_3$ melt maintained at high temperatures while rotating it. However, there is a lithium shortage in the melt at 1250 °C. Therefore, a single crystal obtained using the Czochralski method does not have a Li 50% and Nb 50% stoichiometric composition but a nonstoichiometric composition with Li shortage. As shown by the dashed line in Fig. 4.3, the Curie point (the ferroelectric-paraelectric phase-transition temperature) of $LiNbO_3$ is sensitive to the Li_2O constituent and changes with composition from 1020 to 1180 °C. The important optical property, the refractive index, changes depending on composition. Further, since this is a nonstoichiometric composition, defects occur within the $LiNbO_3$ crystal, and $LiNbO_3$ crystals can easily undergo optical damage from laser light, etc., used in optical applications. Thus, $LiNbO_3$ with a good stoichiometric composition is desired.

While it is difficult to obtain a stoichiometric composition $LiNbO_3$ using the commonly used Czochralski method, stoichiometric $LiNbO_3$ can be synthesized using the chemical solution method.

The chemical solution method can be used to synthesize an epitaxial thin film, which is also considered to be a single-crystalline film or preferentially oriented film, where the crystal growth surface is oriented in a specific direction.

4.3.2.2 Synthesis and Structure of $LiNbO_3$ (LN) Precursor

For thin film synthesis, a precursor solution containing the component element compounds is coated onto a substrate to prepare the precursor thin film. The following coating methods exist among others: the dip-coating method, in which the precursor film is created by immersing the substrate in the solution and raising it at a fixed speed; the spin-coating method, in which solution droplets are coated onto the substrate via rotation at high speeds; and the spray-coating method, in which the solution is spray-coated onto the substrate. A variety of materials are used for the substrate depending on its purpose and the application in which it will be used; substrates include a metal (platinum, etc.), semiconductor (silicon), silica glass, an insulator (oxide single crystal, etc.), and so on.

When synthesizing an epitaxial thin film, an appropriate single-crystal substrate must be selected with consideration given to lattice matching with the target crystal. Frequently used single-crystal substrates are magnesia and sapphire (α-alumina).

Figure 4.4 shows the synthesis procedure for a LN precursor complex alkoxide, as well as LN powder and thin film synthesis processes [23]. Equimolar lithium ethoxide ($LiOC_2H_5$ or LiOEt) and niobium ethoxide ($Nb(OC_2H_5)_5$ or $Nb(OEt)_5$) are heated to reflux in anhydrous ethanol and made to react for a specified amount of time. Water (1.0 equivalent to LN precursor) to the precursor complex alkoxide diluted with ethanol is added dropwise at room temperature. This is again heated to reflux for 24 h, and a homogeneous solution is prepared. The synthesized precursor solution is concentrated to a concentration of 0.1 mol L^{-1}. The solution is hydrolyzed, and after solvent elimination and drying, heat treatment is performed at a specified temperature resulting in the preparation of LN powder.

For precursor film formation, dip coating is performed. The substrate is immersed in LN precursor solution, and after raising the substrate at a fixed speed, it is dried to form the precursor film on the substrate. It is then crystallized under controlled conditions to obtain the LN thin film.

Figure 4.5 shows the ^{93}Nb nuclear magnetic resonance spectrum of the starting material niobium ethoxide and the LN precursor measured in ethanol. Within the solution, niobium ethoxide has become an equilibrium mixture of the monomer and dimer (Fig. 4.6a) [24] and shows two resonance lines at −1067 ppm and −1305 ppm (Fig. 4.5a). Meanwhile, the LN precursor—synthesized by reacting niobium ethoxide and lithium ethoxide in ethanol at 80 °C for 24 h—shows only a single resonance at −1156 ppm; its chemical shift also differs from that of niobium ethoxide (Fig. 4.5b). From these facts, it is known that the LN precursor has

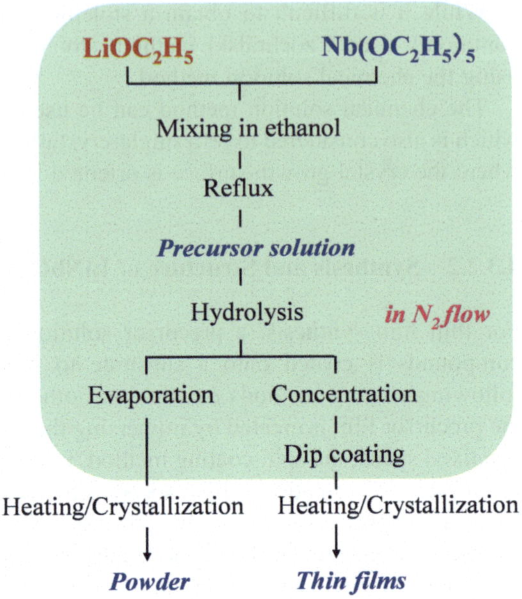

Fig. 4.4 Synthesis procedure for LN precursor, LN powder, and LN thin film

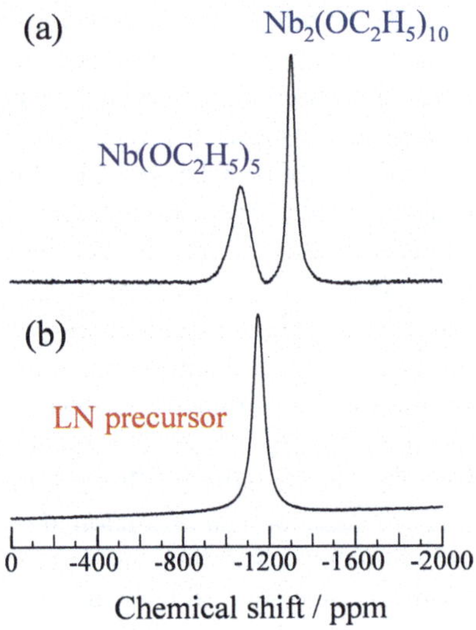

Fig. 4.5 ^{93}Nb nuclear magnetic resonance spectrum of the starting material niobium ethoxide ($Nb(OC_2H_5)_5$) and the LN precursor measured in ethanol (**a**) niobium ethoxide, (**b**) LN precursor

structures including the $Nb(OEt)_6$ octahedral structure, as shown in Fig. 4.6b. Further, from the ^{13}C NMR results, the LN precursor contains only one type of ethoxy group, which is attributed to Nb-OC_2H_5. This LN precursor takes a structure in the ethanol solution such as that shown by Li[$Nb(OEt)_6$]. Measurement of these NMR spectra makes it easy to know whether or not the LN precursor has been generated. When composition deviation exists and the precursor has not been sufficiently generated, different resonance lines are observed in the ^{93}Nb NMR spectrum. From the precursor solution that has not taken the desired configuration

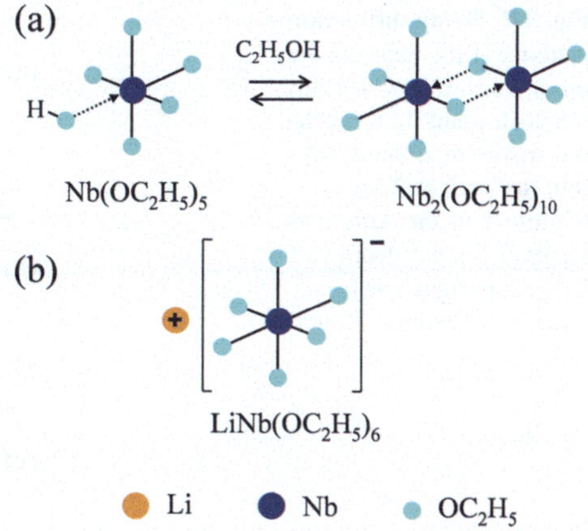

Fig. 4.6 Molecular structures of niobium ethoxide and LN precursor in ethanol (**a**) niobium ethoxide, (**b**) LN precursor. (Reprinted with permission from Ref. [24]. Copyright 1995 John Wiley & Sons)

within the process of crystallization to $LiNbO_3$, a second phase is generated as described below.

4.3.2.3 Low Temperature Synthesis of $LiNbO_3$ (LN) Powder

Powder obtained from the hydrolysis of product formed from the LN precursor was heat-treated in air for 2 h at 350 °C, and the obtained product was analyzed using X-ray diffraction; results are shown in Fig. 4.7. In the precursor immediately after mixing, there is still insufficient reaction of the lithium ethoxide and the niobium ethoxide, and the powder after heat treatment is a mixture of $LiNbO_3$, $LiNb_3O_8$, and Li_3NbO_4. Even after 4 h of reaction, although $LiNbO_3$ is the main phase, it is still a mixture. From the precursor synthesized for 22 h by heating to reflux, the powder after heat treatment at 350 °C crystallized to single-phase $LiNbO_3$, as shown in Fig. 4.7c.

When a LN precursor that has undergone hydrolysis with a controlled quantity of water is used, the LN crystallization temperature can be further reduced. During the hydrolysis stage of Fig. 4.4, 1.0 equivalent water is added. When this has been heated to reflux at 80 °C for 24 h, a homogenous precursor solution is obtained. The powder obtained from this precursor is amorphous, as shown in Fig. 4.8b. When this powder is crystallized in a water vapor/oxygen mixture flow, a crystalline LN powder is obtained at 250 °C, as shown in Fig. 4.8a. By using an appropriately configured precursor, and with decomposition of the organic matter in a controlled environment, crystalline LN can be synthesized at temperatures lower than when using conventional processes.

4.3.2.4 Formation of Stoichiometric Epitaxial $LiNbO_3$ (LN) Thin Film

LN crystal is trigonal with hexagonal closest packing of the oxide ions. Thus, when sapphire (α-Al_2O_3) single crystal, also having hexagonal closest packing of oxide ions, is used as the substrate, oriented thin films and epitaxial thin films can be

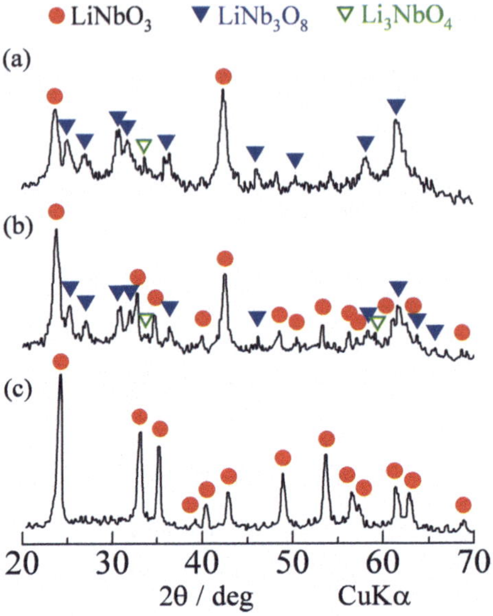

Fig. 4.7 X-ray diffraction patterns of the heat-treated products at 350 °C for the 0 h-, 4 h-, and 22 h-reacted precursors in ethanol, (**a**) 0 h, (**b**) 4 h, (**c**) 22 h (Courtesy of the American Ceramic Society from Ref. [23]. Copyright 1987 American Ceramic Society)

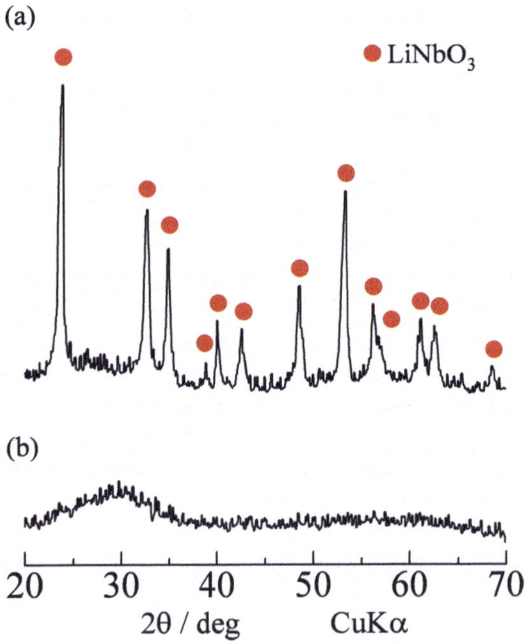

Fig. 4.8 X-ray diffraction patterns of the as-prepared powder from the LN precursor and the heat-treated product at 250 °C in a water gas/oxygen mixture flow (**a**) heat-treated product at 250 °C, (**b**) as-prepared powder (Courtesy of the American Ceramic Society from Ref. [23]. Copyright 1987 American Ceramic Society)

synthesized. There are three types of sapphire substrate: (012) plane, (110) plane, and (001) plane; oriented thin films can be produced on any of these substrates. Figure 4.9 shows X-ray diffraction patterns of LN film crystallized at 400 °C with silicon (100) and each sapphire plane as the substrates. As shown in Fig. 4.9a, a LN

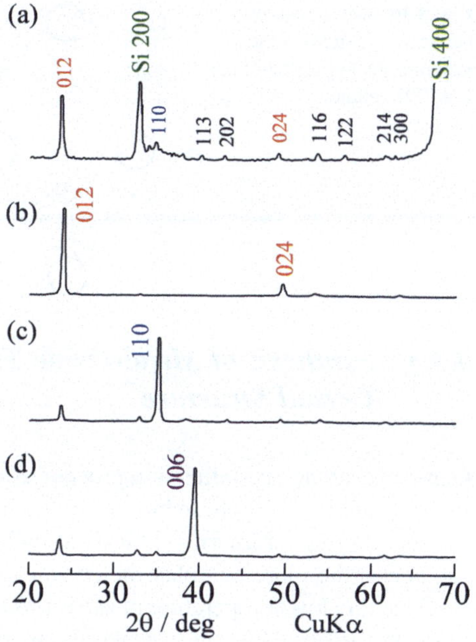

Fig. 4.9 X-ray diffraction patterns of LN film crystallized at 400 °C with silicon (100) and each sapphire plane as the substrates, (**a**) Si (100), (**b**) sapphire (012), (**c**) sapphire (110), (**d**) sapphire (001)

thin film on a silicon substrate shows the same diffraction pattern as a polycrystalline powder. Meanwhile, only the strong diffraction lines of LN 012, 011, and 006 are observed for the sapphire (012) plane, (110) plane, and (001) plane, respectively. We thus know that LN oriented in these crystal planes has been produced on these sapphire substrates. When LN thin films crystallized on sapphire (001) are further analyzed using the X-ray pole figure method, three-fold symmetry spots are displayed, showing that this is an epitaxial film. That is, while the LN crystal lattice has three-dimensional regularity vis-à-vis the sapphire (001) plane of the substrate, the LN is also a single-crystalline film.

Figure 4.10 shows the relationships between ion sites on the sapphire (001) plane and the LN (001) plane. Since LN has a larger lattice constant, the open circles located on the large hexagonal shape show oxide ions. The lattice mismatch between the sapphire and LN is 8.20% in the a-axis direction and 6.67% in the c-axis direction. The refractive indices of the epitaxial LN thin film synthesized on the respective sapphire planes were as follows: 2.308 on the (012) plane, 2.314 on the (110) plane, and 2.386 on the (001) plane; these are appropriate values that show the stoichiometric composition of the LN.

The LN film on the sapphire (001) plane showed the high refractive index of 2.386. This is thought to be because the LN thin film undergoes compressive stress from within the (001) plane of the sapphire substrate due to differences in their coefficients of thermal expansion. Cases of thin film properties being impacted by stress from the substrate are well known. An example is the Curie point shift and increase of coercive filed of $BaTiO_3$ films because of high compressive stresses (≧ 400 MPa) [25]. Of deep interest is the use of stress to control thin film properties.

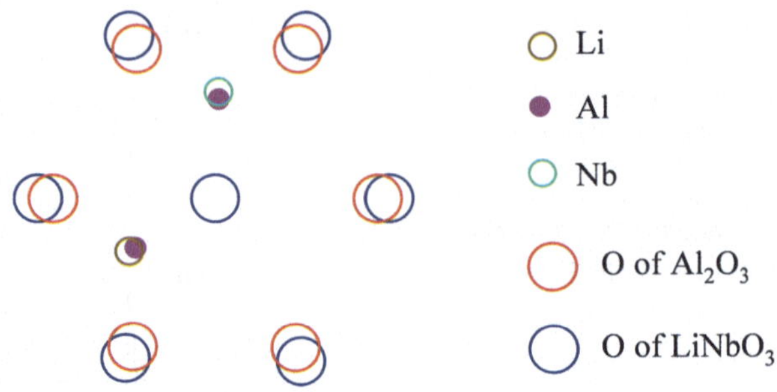

Fig. 4.10 Relationships between ion sites on the sapphire (001) plane and the LN (001) plane

4.3.3 Synthesis of Metal-Oxide Thin Films Having Complex Crystal Structure

Generally, in the crystallization process using the chemical solution method, it is not in the desired crystal phase but in a low-temperature phase (metastable phase) that is not reported for ordinary synthesis at high temperatures when prior crystallization can occur; the more complex the crystal structure, the more often this trend occurs.

Once this low-temperature phase is produced, heat treatment at high temperatures must be performed again to make the transition to the target crystal phase. At this time, due to reactions of the thin film and substrate, a second phase is produced, and abnormal grain growth occurs. The result is a decline in the translucency and surface smoothness of the thin film; nonuniform particle size distribution occurs, and one can no longer obtain a good-quality epitaxial film having the desired optical and electrical properties. This is the case for $Ba_2NaNb_5O_{15}$ (BNN) thin films having a tungsten bronze structure. BNN is a ferroelectric and a crystal that displays electro-optical effects. As the precursor film crystallizes, a crystal hexagonal structure is generated at low temperatures. BNN with this structure has inferior dielectric properties and shows no electro-optical effects. In this case, one uses a thin underlayer film crystallized beforehand with the tungsten bronze structure; on this, a precursor film of ordinary thickness is formed, and crystallization is performed by rapid heating. In this way, an epitaxial BNN thin film displaying dielectric hysteresis and second harmonic generation can be obtained [21, 26].

Recently, there has been interest in bismuth-layered perovskite structure oxide ($Bi_4Ti_3O_{12}$) as a nonvolatile ferroelectric memory material. Compared with the Pb $(Zr,Ti)O_3$ thin film, oxygen defects do not readily occur even when fabricated on a platinum electrode substrate, and its fatigue properties are superior when there are repeated polarization reversals. As shown in Fig. 4.11, this oxide has a complex crystal structure in which a perovskite lattice is sandwiched between bismuth oxide layers. For application in devices, it will be necessary to form the thin film material on a substrate (like semiconductor silicon) onto which the platinum electrode has been deposited. In this case, crystallization must be performed at temperatures of 650 °C or below to prevent reactions between the thin film and the substrate. Layered compound thin films having this complex composition and structure can also be synthesized at 650 °C or below using the chemical solution method; the favorable

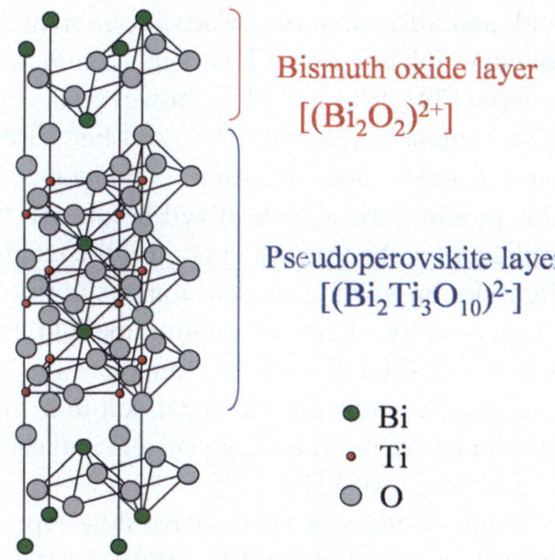

Fig. 4.11 Structure of layered perovskite, $Bi_4Ti_3O_{12}$

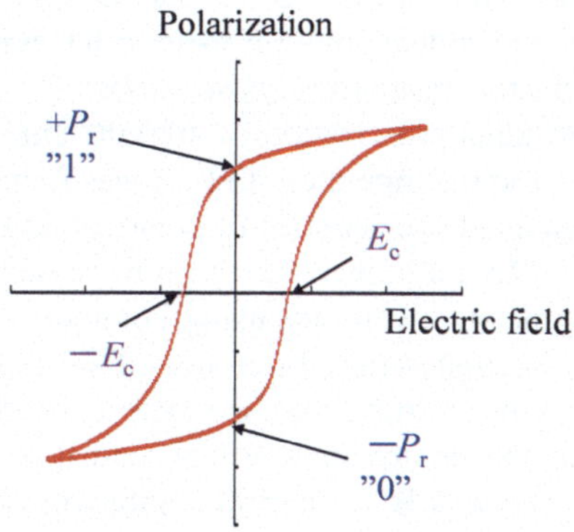

Fig. 4.12 Polarization versus electric field hysteresis curve for $Bi_4Ti_3O_{12}$ thin film

hysteresis characteristics of a $(Bi,Nd)Ti_3O_{12}$ thin film are shown in Fig. 4.12 [27]. The residual polarization value (P_r) of this film is 21.6 μC cm^{-2}, and its coercive electric field (Ec) is 70 kV cm^{-1}. These $+P_r$ and $-P_r$ correspond to "1" and "0," respectively. Also, uniform Ge trace doping utilizing the characteristics of the solution method is applied to crystallize a Ge-doped $(Bi,Nd)Ti_3O_{12}$ thin film with good properties even at 600 °C [28].

4.3.4 Synthesis and Applications of Multiferroic Thin Film

Substances with combined properties of ferroelectricity, ferromagnetism, ferroelasticity, etc. are called "multiferroic" and are attracting attention as new functional materials. Bismuth ferrite ($BiFeO_3$, BF) is a perovskite-type oxide and is a multiferroic material that shows superior ferroelectricity and antiferromagnetism in thin films at room temperature. Since the magnetization direction in an electric field and direction of polarization in a magnetic field of BF are controllable, abundant related research is being conducted for BF as a next-generation electronic material [29]. With BF films, however, due to Bi volatilization, Fe ions easily attain a mixed valence state with Fe^{2+} and Fe^{3+}. Reduction of Fe^{3+} to Fe^{2+} results in oxide ion vacancies, and electronic conductance occurs, increasing the leakage current. This problem must be resolved before BF films can be used in ferroelectric device applications. An attempt to control the occurrence of oxide ion vacancies involves a slight doping with aliovalent ions. Addition of even trace levels of such elements is effective in the chemical solution method. For example, when Fe ions located at site A are substituted with Mn and Cr ions at just a few mol%, BF film can be synthesized using the chemical solution method. It has been reported that such a BF film has reduced leakage current and shows good ferroelectric hysteresis at room temperature [30, 31].

Solid solution of other perovskite-type oxides in BF enables the synthesis of perovskite single-phase BF-ABO ferroelectric thin films with improved insulation properties. It has been reported that 0.7BF-0.3$PbTiO_3$ (PT) thin film synthesized by the chemical solution method has a reduced leakage current, with improved ferroelectric properties. Also, 0.7BF-0.3 PT film doped with 5 mol% Mn showed a residual polarization of 40.0 $\mu C\ cm^{-2}$ and a coercive electric field of 100 $kV\ cm^{-1}$ at room temperature [32]. It was found that the decreased coercive electric field also enabled easy control of polarization reversal.

Recently, there has been increased interest in energy-harvesting materials, which generate electricity using commonly existing phenomena such as vibration, light, and temperature differences. BF has a comparatively narrow band gap, a shortcoming in a ferroelectric. Because it absorbs visible light, a photocurrent is generated due to the photovoltaic effect, and this photocurrent can be converted into electrical energy. A BF film (500 nm thickness) synthesized on a substrate using the chemical solution method was irradiated with 450 nm wavelength light [33]. The results are shown in Fig. 4.13a. A photocurrent with extremely good responsiveness to on-off light irradiation was detected. The same experiment was performed with a laminated film made with 1 mol% Al-doped ZnO film (35 nm thickness) on BF film (300 nm thickness). As shown in Fig. 4.13b, there is major improvement in the photocurrent value. This is thought to be due to a combination of the photocurrent generated at the pn junction interface between the p-type BF film and the n-type Al-doped ZnO film and the self-bias current within the BF film. This generated photoinduced current showed high stability, continuing for 24 h.

BF is also being investigated as a vibration power generation element for thin film and microelectromechanical system (MEMS) applications. Novel properties are still being discovered, and it is a deeply interesting material.

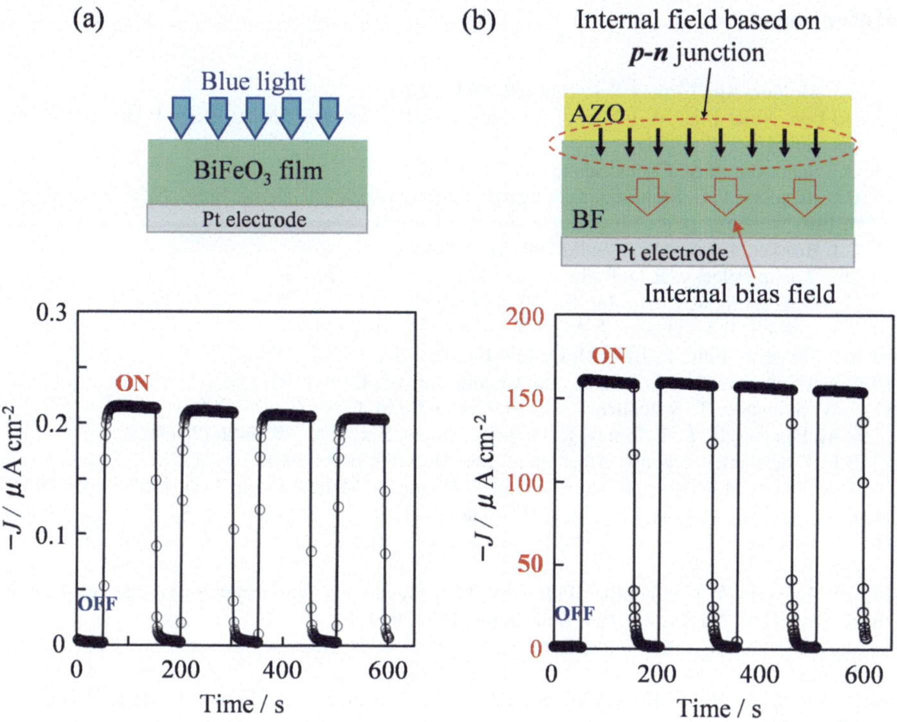

Fig. 4.13 Zero-bias photoelectric current properties (measured at room temperature) of unpoled (**a**) $BiFeO_3$ single-layer and (**b**) 1 mol% Al-doped ZnO/$BiFeO_3$-layered thin films on Pt/TiOx/SiO_2/Si substrates as a function of time under dark (OFF) and blue light (λ = 450 nm, 1.0 mW/cm^2) irradiation (ON) conditions (Copyright 2015 The Japan Society of Applied Physics)

4.4 Current Trend of Ceramic Synthesis

In the synthesis of functional inorganic material powders and thin films, it is necessary to control the crystal structure and chemical composition. Yet, the control of microstructure is the key to property manifestation. In powders, these properties are structure, particle shape, particle size (diameter), and particle size distribution. In thin films, these properties include post-crystallization particle size (diameter), particle size distribution, surface smoothness, etc. The same holds for nanoparticles and nanomaterials that are of recent particular interest, and the chemical solution method is one important technique for the synthesis of these new materials.

References

1. B.J. Mulder, Am. Ceram. Soc. Bull. **49**, 990 (1970)
2. G.J.D.A. Soler-Illia, M. Jobbagy, R.J. Candal, A.E. Regazzoni, M.A. Blesa, J. Dispersion Sci. Technol. **19**, 207 (1988)
3. G. Johansson, Acta. Chem. Scand. **14**, 771 (1960)
4. D.C. Bradley, R.C. Mehrotra, D.P. Gauer, *Metal Alkoxide* (Academic Press, New York, 1978), p. 149
5. C.J. Brinker, J. Non-Cryst. Solid **100**, 31 (1988)
6. D.L. Leslie-Pelecky, R.D. Rieke, Chem. Mater. **8**, 1770 (1996)
7. H. Weller, Angew. Chem. Int. Ed. **32**, 41 (1993)
8. V.K. LaMer, R.M. Dinger, J. Am. Chem. Soc. **72**, 4847 (1950)
9. W. Stöber, A. Fink, E. Bohn, J. Colloid Interface Sci. **26**, 62 (1968)
10. E.A. Barringer, H.K. Bowen, J. Am. Ceram. Soc. **65**, C199 (1982)
11. R.W. Schwartz, T. Schneller, R. Waser, Comp. Rend. Chim. **7**, 433 (2004)
12. R.M. Pasquarelli, D.S. Ginley, R. O'Hayre, Chem. Soc. Rev. **40**, 5406 (2011)
13. B.E. Yoldas, D.P. Oatlow, Am. Ceram. Soc. Bull. **59**, 640 (1980)
14. T. Kumagai, H. Yokota, K. Kawaguchi, W. Kondo, S. Mizuta, Chem. Lett. **1987**, 1645 (1987)
15. F. Geotti-Bianchini, M. Guglielmi, P. Polato, G.D. Soraru, J. Non-Cryst. Solid **63**, 251 (1984)
16. A. Makishima, H. Kubo, K. Wada, Y. Kitami, T. Shimohira, J. Am. Ceram. Soc. **69**, C127 (1986)
17. E.R. La Serra, Y. Charbouillot, P. Baudry, M.A. Aegerter, J. Non-Cryst. Solids **121**, 323 (1990)
18. H. Dislich, E. Hussmann, Thin Solid Films **77**, 129 (1981)
19. S. Hirano, T. Yogo, K. Kikuta, K. Noda, M. Ichida, A. Nakamura, J. Am. Ceram. Soc. **72**, 2956 (1995)
20. T. Yogo, K. Niwa, K. Kikuta, M. Ichida, A. Nakamura, S. Hirano, J. Mater. Chem. **7**, 929 (1997)
21. T. Yogo, W. Sakamoto, T. Isaji, M. Ichida, A. Nakamura, S. Hirano, J. Am, Ceram. Soc. **82**, 2672 (1999)
22. J.R. Carruthers, G.E. Peterson, M. Grasso, P.M. Bridenbaugh, J. Appl. Phys. **42**, 1846 (1971)
23. S. Hirano, K. Kato, Adv. Ceram. Mater. **2**, 142 (1987)
24. T. Yogo, K. Kikuta, Y. Ito, S. Hirano, J. Amer. Ceram. Soc. **78**, 2175 (1995)
25. S.B. Desu, Phys. Stat. Solid A **141**, 119 (1994)
26. T. Yogo, W. Sakamoto, T. Isaji, K. Kikuta, S. Hirano, J. Am. Ceram. Soc. **80**, 1767 (1997)
27. W. Sakamoto, Y. Mizutani, N. Iizawa, T. Yogo, T. Hayashi, S. Hirano, Jpn. J. Appl. Phys. **43**, 6599 (2004)
28. W. Sakamoto, Y. Mizutani, N. Iizawa, T. Yogo, T. Hayashi, S. Hirano, Jpn. J. Appl. Phys. **42**, L1384 (2003)
29. G. Catalan, J.F. Scott, Adv. Mater. **21**, 2463 (2009)
30. J.Z. Huang, Y. Wang, Y. Lin, M. Li, C.W. Nan, J. Appl. Phys. **106**, 063911 (2009)
31. J.K. Kim, S.S. Kim, W.J. Kim, A.S. Bhalla, R. Guo, Appl. Phys. Lett. **88**, 132901 (2006)
32. W. Sakamoto, A. Iwata, T. Yogo, J. Appl. Phys. **104**, 104106 (2008)
33. T. Katayama, W. Sakamoto, I. Yuitoo, T. Takeuchi, K. Hayashi, T. Yogo, Jpn. J. Appl. Phys. **54**, 10NA05 (2015)

Chapter 5
Chemical Functions of Ceramics

Shu Yin and Tsugio Sato

Abstract Chemical functions of ceramics are strongly related to the particle size, crystalline plane, morphology, and synthesis methods. In this chapter, the physical and chemical properties of fine ceramics materials are introduced. Environmentally friendly soft chemical processes, including solvothermal/hydrothermal process and mechanochemical process, are the effective methods for the synthesis of various functional materials and mixed anion type visible light-induced photocatalysts. The mixed anion type photocatalytic compounds consisting of N/O, N/F/O, S/O, and N/C/O show excellent visible light absorption ability, indicating the potential applications in environmental purifications with high solar light utilization efficiency. Various functions such as full-spectra active long wavelength light-induced photocatalyst, full-time active photocatalyst system, UV shielding, oxygen storage capacity (OSC), and hydrophilicity of oxide ceramics based on the precise design of components, particle size, and morphology are introduced also.

Keywords Mixed anion · Photocatalyst · Oxygen storage capacity · Hydrophilicity · Multifunctionality

5.1 Physical and Chemical Properties of Ceramics Fine Particles

5.1.1 Surface Energy

Physical and chemical properties of ceramics are greatly affected by their particle size and morphologies. With the decrement of particle size, their specific surface area becomes large, and the effect of the surface energy becomes significant. The surface energy is the energy consumed by the destruction of the intermolecular

S. Yin (✉) · T. Sato
Institute of Multidisciplinary Research for Advanced Materials, Tohoku University, Sendai, Japan
e-mail: yin.shu.b5@tohoku.ac.jp; tsugio.sato.d8@tohoku.ac.jp

J. Hojo (ed.), *Materials Chemistry of Ceramics*,
https://doi.org/10.1007/978-981-13-9935-0_5

chemical bond when the surface of the material is created. Surface atoms have more energy than those inside the material. Therefore, according to the *principle of minimum energy* [1], atoms spontaneously tend to be inside the bulk material rather than on the surface. The surface energy is defined as the excess energy at the surface of a material compared to the bulk. In other words, it is the work required to build an area of a particular surface. The decomposition of a solid material into small pieces requires destruction of its chemical bonds and needs to consume energy. The energy required to break down the material into small pieces is equal to the energy released from the surface of the small piece of material to build the bulk. The newly formed unstable surfaces reduce the surface energy by surface atomic recombination [2].

5.1.2 *Properties of Fine Particles*

The surface energy of the solid is increased by repeating the cleavage of the bond. Therefore, for small particles, the ratio of the surface energy to the total energy increases greatly and becomes significant. The high surface area to volume ratio of nanoparticles provides a huge driving force for diffusion. A bulk material has constant physical properties regardless of its size, but at the nanoscale, the size-dependent properties are often observed, as they are small enough to confine their electrons and produce quantum effects. Some unexpected or unique properties of nanoparticles are related to their large surface area of the material. Nanoparticles are of great interest as they are a bridge between bulk materials and atomic or molecular structures.

5.1.2.1 Vapor Pressure of Fine Particles

It has been shown on theoretical research that, if a liquid is broken up into small droplets or if it is held in pores or other similar spaces which are of capillary dimensions, its vapor pressure is a function of the curvature of the surface [3].

The vapor pressure of liquid particles can be calculated by the Kelvin equation (Eq. 5.1).

$$\ln P/P_0 = 2M\gamma/(RT\rho \bullet r) \tag{5.1}$$

where P is the actual vapor pressure, P_0 is the saturated vapor pressure, γ is the surface energy, M is the molar molecular weight of the liquid, ρ is the density of the liquid, R is the universal gas constant, r is the radius of the droplet, and T is the absolute temperature.

In the case of water drop at 20 °C, the surface energy is 73 erg cm^{-2}, the saturated vapor pressure (P_0) is 2.34 kPa, and the relationship between liquid particle radius and vapor pressure can be expressed in Eq. 5.2 and be plotted as shown in Fig. 5.1.

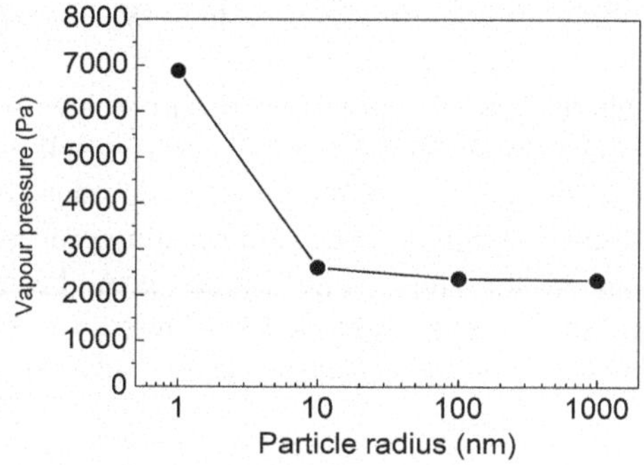

Fig. 5.1 Relationship between particle radius and vapor pressure of water drop at 20 °C

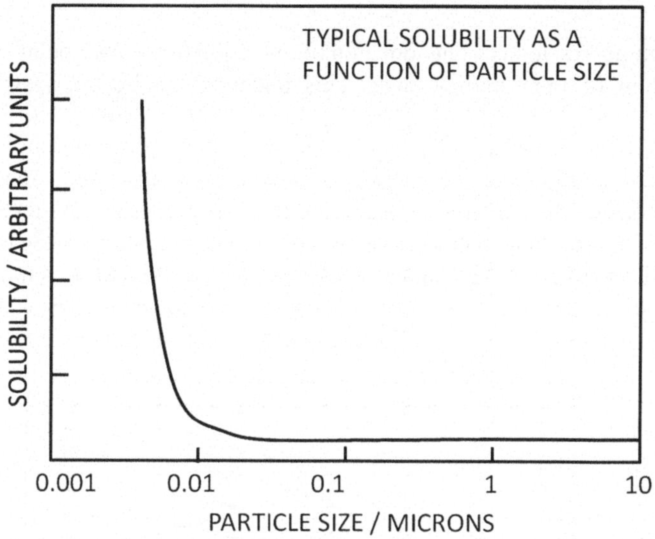

Fig. 5.2 Typical solubility as a function of particle size [6]

$$\ln P/P_0 = 1.08 \times 10^{-9}/r \tag{5.2}$$

When the size is less than 10 nm, the increment of vapor pressure becomes significant.

5.1.2.2 Solubility of Fine Particles

The Gibbs free energy equation shows that solubility is dependent on particle size and their surface area to volume ratio. Figure 5.2 shows the typical solubility as a function of particle size. Similar to the Kelvin equation, the following Eq. 5.3 can be utilized for the calculation of solubility of fine crystallines/particles:

$$\ln S_r/S_\infty = 2M\gamma/(RT\rho \bullet r) \quad (5.3)$$

where S_r is the solubility of particles with radius of r and S_∞ is the solubility of materials with flat surface [4]. Very small particles give too high surface energy, and the solution is unstable long-term because a process called the Ostwald ripening [5] occurs. During the Ostwald ripening process, the small crystals are dissolved easily, but at later stage large crystals grow because they have lower solubility and the solution is supersaturated with respect to the larger crystals. As a result, the small particles are being consumed as the larger ones grow bigger.

5.1.2.3 Melting Point Depression of Fine Particles

Melting point depression is a phenomenon of reduction of the melt point of materials with reduction of their particle size. This phenomenon becomes significant in nanoscale materials, in which the melting point is several hundred degrees lower than that of bulk materials. Changes in melting point occur because nanoscale materials have a much larger surface-to-volume ratio than bulk materials and dramatically affect their thermodynamic and thermal properties. As an example, it is found that the melting temperature of GaN ceramic nanoparticles consistently decreases with the size of the nanoparticle [7] (Fig. 5.3). Further, it is found that the particle shape can also affect the melting temperature of nanoparticles, and this effect on the melting temperature becomes larger with decreasing particle size. Another

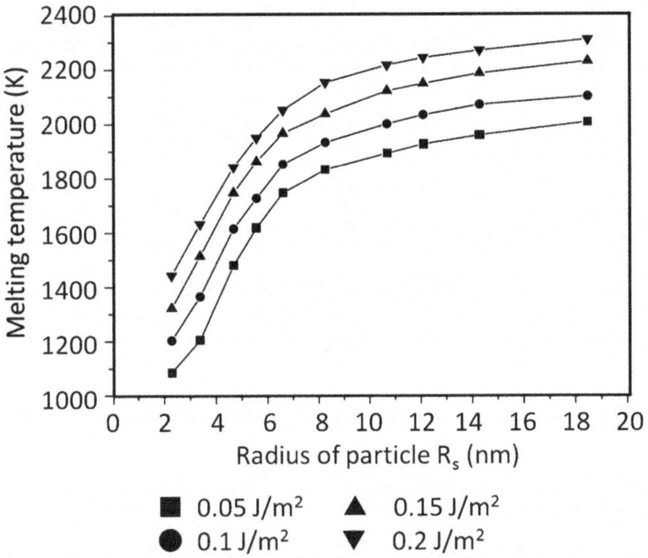

Fig. 5.3 Variation of melting point of GaN nanoparticles against the size of the particles. Reprinted from Ref. [7] with open access from Hindawi. The particles with different surface energy are shown with various symbols

example is gold nanoparticles. Gold nanoparticles with 2.5 nm size melt at much lower temperatures (about 300 °C) than the gold bulk (1064 °C) [8].

5.1.2.4 Chemical Reactivity of Fine Particles

Specific surface area has a great effect on the reactivity of materials. The larger surface provides larger adsorbability and intimate contact among the substances and results to higher chemical reaction among each other. The specific surface area is increased as the particle size becomes smaller. The specific surface area is also increased if the particle has pores. Even if the same material has the same weight and volume, the surface activity and adsorbability change greatly according to their specific surface area.

It is accepted that the specific surface area (SSA) of nonporous spheres can be calculated from the particle size by the following Eq. 5.4:

$$SSA(\mathrm{BET}) = 6/(\rho \cdot D) \tag{5.4}$$

where SSA [m^2/g] is the specific surface area measured by BET, ρ is the density of particle, and D is the particle diameter.

5.2 Photocatalytic Functions of Ceramics

It is known that photon energy is solely a function of the photon's wavelength ($E = h\upsilon$). The photon with longer wavelength possesses the lower energy. Normally, after irradiation of light on the semiconductor, the electron-hole pairs are produced on the surface. Based on the light-induced process, as shown in Fig. 5.4, many applications such as photocatalyst, solar cell, optical sensor, UV shielding, and luminescence properties can be realized by utilizing the light-induced process or the chemical activities of the electrons or holes [9, 10].

Photocatalytic reactions are using sunlight as energy source to excite semiconductor. It is one of the promising processes which do not use fossil energy. Since the discovery of photocatalysis phenomenon in 1972 [11], the photocatalytic functions of ceramics have attracted widespread attentions of researchers and engineers.

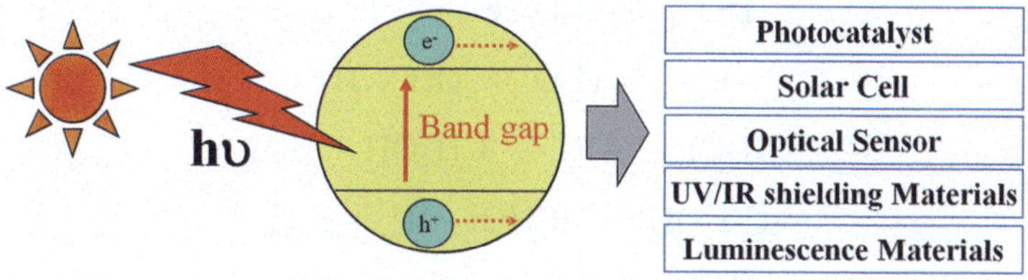

Fig. 5.4 Light-induced process on semiconductor and its applications

Because photocatalytic reaction system utilizes clean and inexhaustible solar energy, it is a low environmental load type chemical process and expected to have great potential in applications on the purification of air and wastewater, organic degradation, hydrogen production, and solar cell [9–15]. Until now, the titanium oxide photocatalyst is regarded as one of the most effective photocatalysts because of its excellent stability, non-toxicity, low expense, and high activity. Besides titanium oxide, some oxides such as zinc oxide, tungsten oxide, and ABO_3-type perovskite compounds are well studied also. It is well known that the photocatalytic activity is closely related to the crystallinity, phase composition, BET-specific surface area, micro/nano structure, and morphology of the materials.

5.2.1 Photocatalytic Activity of Titania Ceramics

Titania is the principal white pigment in the world and has wide applications in cosmetics, catalyst, and catalyst carrier because of its excellent physical and chemical properties [13, 14]. In 1972, Fujishima and Honda [11] discovered the photocatalytic splitting of water on TiO_2 electrodes. Until now, the titanium oxide photocatalyst is still one of the most excellent and effective photocatalysts in the practical applications.

It is accepted that TiO_2 particles absorb UV light to generate electron/hole pairs (Eq. 5.5). The electrons are photo-induced to conduction band ($e^-{}_{CB}$), while the holes in valence band ($h^+{}_{VB}$) are subsequently trapped by H_2O molecule to form H^+ and •OH radical (Eq. 5.6).

Figure 5.5 illustrates the energetic diagram and photo-induced charge transfer process in a photocatalytic TiO_2 particle. Although the bandgap of titania (3.2 eV) is larger than the potential energy for electrochemical decomposition of water (1.23 eV), pure water is difficult to be photo-decomposed on titania surface because of the rapidly occurred backward reaction of H_2 and O_2. In the absence of oxygen and in the presence of sacrificial species such as methanol, the holes oxidize methanol to produce HCHO, etc., while electrons in the conduction band simultaneously reduce water to form gaseous H_2 as shown by Eqs. 5.6–5.9 [15–18]. These reactions proceed competitively with the recombination of the photo-induced electrons and holes.

$$TiO_2 \overset{h\nu}{\longleftrightarrow} TiO_2(e^-{}_{CB} + h^+{}_{VB}) \tag{5.5}$$

$$h^+{}_{VB} + H_2O \rightarrow \cdot OH + H^+ \tag{5.6}$$

$$CH_3OH + \cdot OH \rightarrow \cdot CH_2OH + H_2O \tag{5.7}$$

$$\cdot CH_2OH \rightarrow HCHO + H^+ + e^-{}_{CB} \tag{5.8}$$

$$2H_2O + 2\,e^-{}_{CB} \rightarrow H_2 \uparrow + 2OH^- \tag{5.9}$$

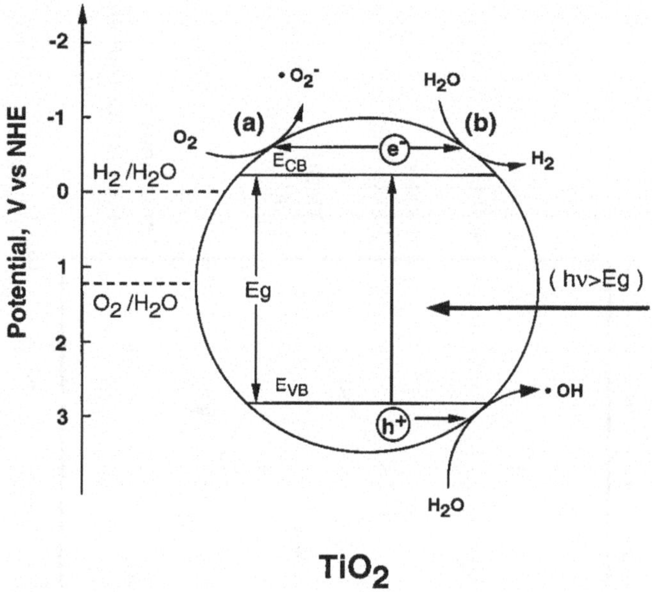

Fig. 5.5 Schematic illustration of the energy correlation and the redox mechanism on titania catalyst surface (**a**) with and (**b**) without oxygen. (Reprinted with permission from Ref. [15]. Copyright 2000 ACS)

$$e^{-}{}_{CB} + O_2 \rightarrow \cdot O_2{}^{-} \tag{5.10}$$

$$\cdot O_2{}^{-} + H^{+} \rightarrow \cdot OOH \tag{5.11}$$

$$C_6H_5OH + \{\cdot OH, \bullet OOH, \text{and/or } O_2\} \rightarrow \text{Intermediates} \rightarrow CO_2 + H_2O \tag{5.12}$$

$$NO + \{\cdot OH, \cdot OOH, \text{and/or } O_2\} \rightarrow HNO_2 \text{ and/or } HNO_3 \quad (80\%) \tag{5.13}$$

$$NO + \{\cdot OH, \cdot OOH, \text{and/or } O_2\} \rightarrow N_2 + O_2 \quad (20\%) \tag{5.14}$$

Meanwhile in the presence of oxygen, the electrons in conduction band are rapidly trapped by the molecular oxygen to form $O_2{}^{-}$ (Eq. 5.10) and then generate highly active ·OOH radicals (Eqs. 5.10 and 5.11) [15, 19]. The environmental pollutant such as phenol in the solution reacts with these reactive oxygen radicals, and/or molecular oxygen, to produce intermediates or complete decomposition products, CO_2 and H_2O [19, 20] (Eq. 5.12). In the case of gas phase reaction, environmental pollutant such as nitrogen monoxide (NO) reacts with above reactive oxygen radicals to produce nitrate acid/nitric acid or nitrogen and oxygen directly [21] (Eqs. 5.13 and 5.14). The produced acid is thought to be neutralized by basic ash existing in air and finally leads to the environmental purification.

If the crystallinity of the photocatalyst increases, the electron-hole recombination rate decreases; consequently, the fluorescence lifetime increases. Figure 5.6 illustrates the relationship among specific surface area, fluorescence lifetime, and hydrogen evolution rate. It is obvious that although amorphous titania possesses large-specific surface area, the photocatalytic activity is quite poor, because of their short

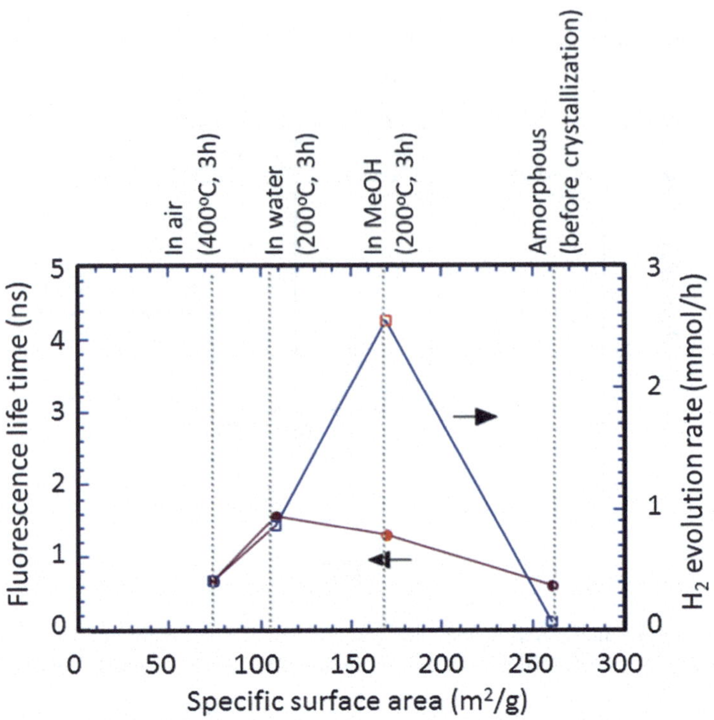

Fig. 5.6 Photocatalytic activities, fluorescence lifetimes, and specific surface areas of titania crystallized under various conditions. (Reprinted from the Ref. [22] with permission from Cambridge University Press)

electron and hole's lifetime, relating to the large amount of crystalline defects which act as the electron-hole recombination center. On the other hand, the titania crystalline, especially crystallized in methanol, shows the highest photocatalytic activity for hydrogen evolution. In addition, the activity of anatase seems to increase with increasing specific surface area. Also, anatase crystallized in water and methanol shows longer fluorescence lifetime than that crystallized in air. These results suggest that the solvothermal reaction results to the formation of fine crystals of titania with perfect crystallinity and results to the excellent photocatalytic H_2 evolution activity.

5.2.2 Design Guideline for High-Active Photocatalytic Materials

According to the photocatalytic reaction mechanism mentioned in Fig. 5.5, it is suggested that the following factors are considered to promote the photocatalysis efficiency for ceramics semiconductor materials (see Table 5.1 and Fig. 5.7).

In order to improve the photocatalytic activity, some effective ways are suggested to enhance the charge transfer or charge separation. High crystallinity and high

Table 5.1 Design guideline for high-active photocatalytic ceramics

No.	Factor	Detail actions	References
1	High crystallinity	Decrease the crystal defects and depress the recombination between photo-induced electrons and holes	[15, 22, 23]
2	High specific surface area	Lead to ultrafine particles and improved adsorption property	[15, 23, 24]
3	Dye-sensitization	Improve the light absorption	[25–27]
4	Narrowing of bandgap	Be realized by colorizing or mixed ions doping	[23, 28, 29]
5	Loading of noble metal	Promote the charge transfer	[27, 30, 31]
6	Utilization of layered compounds	Interlayer space as reaction field to promote charge transfer	[32, 33]
7	Multicomponents and Z-scheme	Combine various components with different bandgap, resulting in the enhancement of charge transfer/separation	[34–36]

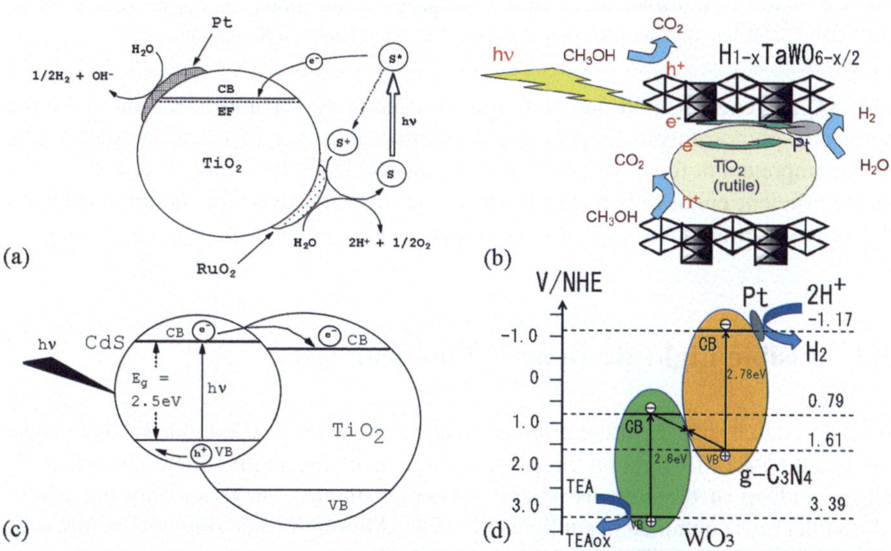

Fig. 5.7 Schematic illustration for the enhancement of charge transfer in various photocatalytic reaction systems (**a**) Dye-sensitization and noble metal loading (S, sensitizer, such as Ru (bpy)); (**b**) utilization of layered compounds; (**c**) multicomponents (heterostructure); (**d**) Z-scheme and noble metal loading. (Reprinted from the Ref. [27, 34, 36] with permission from ACS, RSC)

specific surface area are the two basic important factors to realize the excellent photocatalysis effect, because less crystalline defects and more adsorption sites on the surface of semiconductors are beneficial to depress the electron-hole recombination and promote adsorption of chemicals on the surface [15, 22–24]. Meanwhile in order to improve the light adsorption, some strategies such as dye-sensitization [25–27] and ions doping [23, 28, 29] are normally used recently. Also the loading of noble metal [27, 30, 31] and the utilization of interlayer space of layered compound

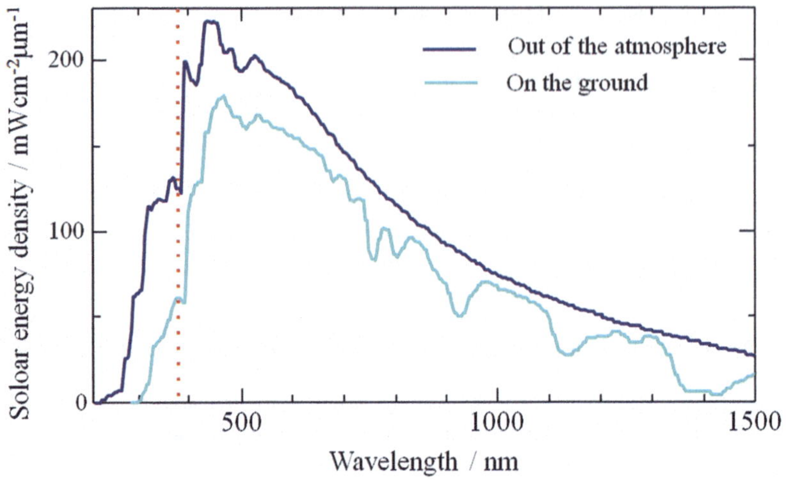

Fig. 5.8 Wavelength distribution of solar light spectra on the ground and at the outside of the atmosphere. Dot line indicates the bound between UV and visible light region

[32, 33] as reaction field are preferred to effectively separate electrons from the surface of photocatalysts. Especially, the formation of heterostructures, such as the multicomponent or the Z-scheme [34–36], can accelerate the charge transfer, leading to the efficient enhancement of photocatalytic activity, even for the photocatalytic decomposition of water to produce hydrogen and oxygen.

5.3 Visible Light Responsive Photocatalysts

It is known that solar light spectrum consists of about 5% of UV light, 45% of visible light, and 55% of IR light on the ground of sea level. Normally, due to the relatively large bandgap of titanium oxide and strontium titanate, they can only use 5% of ultraviolet rays contained in sunlight (Fig. 5.8). Although some semiconductors such as tungsten oxide can absorb visible light effectively, its stability is not enough due to the photolysis of itself during light irradiation. A problem in application of anatase titania as photocatalyst is the large bandgap energy, i.e., anatase only shows photocatalytic activities under UV light irradiation at wavelength < 387 nm, which corresponds to its bandgap value of 3.2 eV. In order to improve the solar light utilization efficiency, it is necessary and important to develop catalysts which can utilize visible light or even infrared long wavelength light.

5.3.1 Anion Doped Visible Light-Induced Photocatalysts

Recently, mixed anion type photocatalysts have become to be a better choice in the fundamental science research and engineering applications [23, 28, 29, 37]. Even

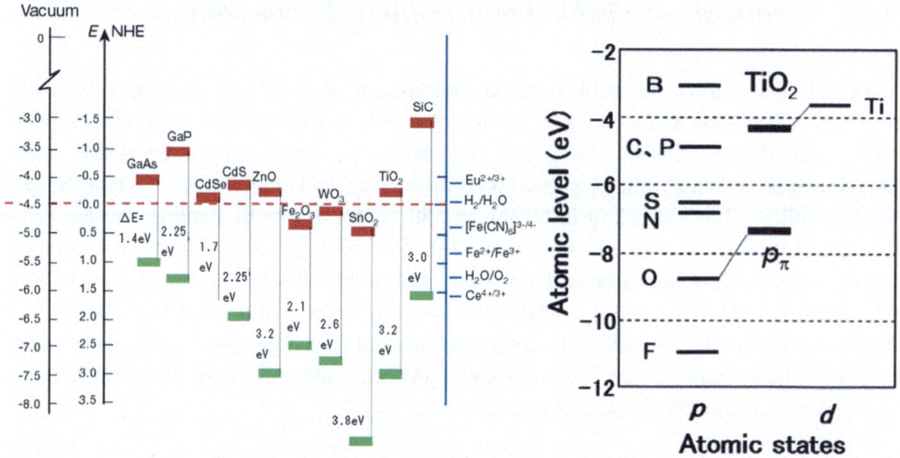

Fig. 5.9 (Left) Band positions of several semiconductors in contact with aqueous electrolyte at pH 1; (Right) comparison of atomic P levels among anions. The bandgap of TiO_2 is formed between O $2p_\pi$ and Ti 3d states. (Reprinted with permission from Ref. [14, 28]. Copyright 2001 Nature Publishing Group and Copyright 2014 ACS)

though the anion elements are closed together in the periodic table, properties such as electronegativity are greatly different. The mixed anion compound, in which multiple anions such as oxygen, sulfur, carbon, nitrogen, and hydrogen are included in the same compound, possesses unique coordination structure and shows the possibility for the innovative functions. The first principle calculation on electronic state density of titanium has been carried out by considering various anion elements doped in oxygen site of titanium oxide. The results imply that the bandgap values can be effectively narrowed by introducing a series of anion elements [28, 29]. Figure 5.9 shows the band positions of several semiconductors and the comparison of atomic P levels among anions. It can be seen that titania has a great potential for the photocatalysis because of its suitable band positions for redox reaction under light irradiation [10], and the bandgap of TiO_2 is formed between $O2p_\pi$ and Ti3d states. It is expected that nitrogen-doped titanium oxide exhibits visible light responsiveness and excellent visible light-induced photocatalytic activity [28].

Various novel anion type photocatalysts have been successfully synthesized and been reported as high-active hydrogen evolution photocatalysts. It is reported that TaON, $LaMg_xTa_{1-x}O_{1+3x}N_{2-3x}$ ($x \geqslant 1/3$) [38], and N/S/C/P monodoping or N–N/C–S/P–P/N–P co-doping $NaTaO_3$ [9] show excellent hydrogen evolution ability based on water splitting. Another example is the solid solution of $LaTaON_2$–$SrTiO_3$, in which the photocatalytic activities for H_2 and O_2 evolution under visible light are successfully realized by bandgap turning using different fraction of perovskite $SrTiO_3$ [39].

5.3.2 *Synthesis of Visible Light-Induced Photocatalysts*

Various mixed anion type visible light-type photocatalysts have been developed by using a variety of methods and techniques. Different synthesis methods possess related advantages and disadvantages. Nitrogen-doped titanium oxide with high-visible light photocatalytic activities can be prepared by sputtering TiO_2 target in N_2 (40%)–Ar gas mixture followed by annealing in N_2 gas at 550 °C for 4 h or by treating anatase TiO_2 powder in NH_3 (67%)–Ar atmosphere at 600 °C for 3 h [28, 29]. The nitrogen-doped titania shows yellowish color and possesses noticeable absorption less than 500 nm. Chemically modified $TiO_{2-x}C_x$ can be synthesized by flame pyrolysis of Ti metal sheet at 850 °C in a natural gas flame with controlled amount of oxygen [40]. Here, some other recent synthesis technologies, including ion injection, low-temperature plasma treatment, and partial oxidation reaction, are introduced.

5.3.2.1 Ion Injection

Ion injection or ion implantation is a material engineering process by which ions of a material are accelerated in an electrical field or magnetic effect and impacted into the lattice of a solid. This process is a universal process for changing the physical, chemical, or electrical properties of materials. By this method, the metal ion-implanted TiO_2 catalyst enables the absorption of visible light up to a wavelength of 400–550 nm, together with excellent visible light-induced photocatalytic activities. As an example, some metal ions such as Cr and V can be introduced into powdered TiO_2 catalysts, which leads to the modification of the electronic state of titania, resulting in the shift of the absorption into the visible light region [21]. Some anions such as fluorine or nitrogen have also successfully been implanted into the lattice of titania and other photocatalysts.

5.3.2.2 Chemical Vapor Deposition and Plasma Treatment

Besides anion-doped titania, nonstoichiometric titania TiO_{2-x} also possesses excellent visible light absorption and photocatalytic activity. The TiO_{2-x} film can be synthesized by a metal-organic chemical vapor deposition (MOCVD) method [40]. It is found that anatase phase with a large oxygen deficiency in fact acts as an efficient photocatalyst for the decomposition of methylene blue under the irradiation of light wavelength > 400 nm. Nonstoichiometric visible light-active titania TiO_{2-x} can also be prepared by a low-temperature hydrogen plasma treatment at 200–400 °C [41]. The prepared powders possess excellent visible light responsive photocatalytic activities for the oxidation of benzoic acid in liquid phase and the oxidation of 2-propanol in gas phase. In addition, the oxygen vacancies can be introduced by the plasma surface treatment, and different colors of titania such as yellow, gray, blue, and dark blue can be obtained by the precise control of the O/Ti ratio.

5.3.2.3 Partial Oxidation Reaction

The partial oxidation reaction method utilizes carbide/sulfide/nitride as starting precursors. Because the precursors have already possessed the chemical bonding between titanium and anions, it is easy to produce the anion doping state with controllable residual anion doping amount, by partially oxidizing the precursors at a suitable treatment temperature in air atmosphere. For example, the C-doped TiO_2 can be obtained by oxidative annealing of TiC in air/O_2 at 600 °C for 5 h [42]. Similarly, S-doped titania can be produced by the same oxidative annealing of TiS_2 in air at 500–600 °C [43]. The yellowish S-doped TiO_2 can be also obtained by mixing titanium isopropoxide with thiourea and ethanol followed by evaporation and calcination in air at 400–700 °C [44]. However, some disadvantages still exist in the mentioned process, such as high cost of the expensive precursors and difficulty to obtain fine particle size/high specific surface areas of the anion-doped photocatalysts.

As mentioned above, there are many kinds of method for the preparation of visible light responsive photocatalysts. It is well known that photocatalytic activity strongly relates with the physical properties such as crystal phase, particle size, specific surface area, crystallinity, and morphology of ceramics. In some cases, the high-temperature treatment results in the formation of high-temperature stable phases, crystal growth, low-specific surface area, decrement of the anion-doped amount, and consequently the decrement of visible light responsive photocatalytic activities. In order to avoid the above disadvantages, it is suggested to utilize low-temperature soft chemical processes to synthesize high-active photocatalysts. In addition, it is expected to avoid using irritant gases, such as NH_3 and H_2S, and expensive precursors, such as TiN and TiS_2, and to depress the formation of poisonous by-products such as SO_x and NO_x during the photocatalyst preparation process.

5.3.2.4 Mechanochemical Doping

Mechanochemical technology is a low-cost and simple way to synthesize functional materials, in which the operation can be carried out at room temperature. In principle, various ions can be effectively doped in the lattice of crystalline, although some contaminations are introduced. Usually, the mechanical stressing induces the formation of fresh oxygen-rich surface, which results in the electron transfer from O^{2-} ion on oxide surface to other organic substances and, as a result, leads to the destruction of weak bonding in organic substance and the formation of new bonding between oxide and nonmetallic element [45]. In the case of photocatalyst synthesis, the fresh active surfaces are exposed by the milling treatment. It is an effective method for doping anions such as nitrogen and fluorine in the lattice of various photocatalysts around room temperature [24, 46–48].

Carbon doping, carbon/nitrogen co-doping, and sulfur doping are easily realized by grinding TiO_2 with adamantane, hexamethylenetetramine (HMT), or sulfur also [47]. In all the cases, the DRS spectra show excellent light absorption in visible light region (Fig. 5.10). The deNO_x activity increases with the increment of

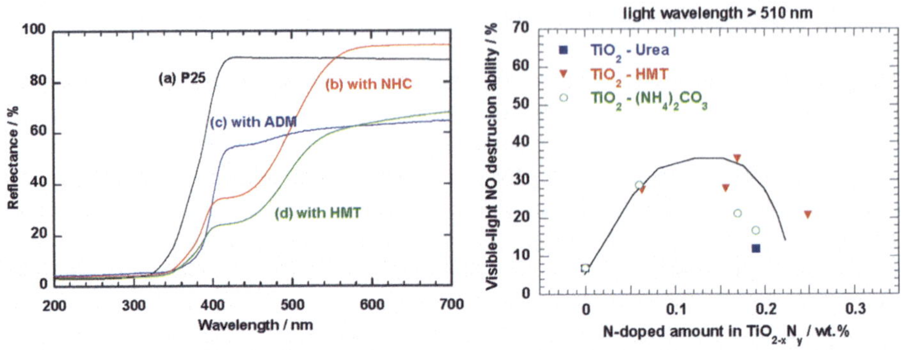

Fig. 5.10 Diffuse reflection spectra (DRS) of the mixed anion type photocatalysts prepared by mechanochemical doping process using urea/ammonia carbonate (NHC), adamantine (ADM), and hexamethylenetetramine (HMT) as reaction reagents and change in DeNOx ability with nitrogen-doped amount. The DRS of P25 is also plotted. (Reproduced from Ref. [48] with permission from Elsevier)

nitrogen-doped amount while decreases in the case of excessive doping. About 0.17 at.% of nitrogen doping results in the highest deNO_x activity under the 510 nm visible light irradiation (Fig. 5.10).

5.3.2.5 Solvothermal Synthesis

Solvothermal reaction is the process that uses high-temperature and/or high-pressure solvents. The dielectric constant and ionic product change depending on the solvent, temperature, and pressure; therefore, the reactivity of solvent changes with acid-base conditions during the reactions. It is possible to utilize the solvothermal reaction to control the nucleation and the crystal growth rate and, finally, produce nano-size crystals with soft agglomeration, and controlling the phase composition or morphology [9, 10, 49].

As shown in Fig. 5.11, the nitrogen-doped titania with anatase, brookite, and rutile single phase can be selectively prepared by the solvothermal process at 190 °C, without any post-solvothermal calcination treatment [23, 50]. Compared with white color commercial titania P25, the as-prepared $TiO_{2-x}N_y$ sample shows violet color. After calcination in air at 200–800 °C for 1 h, it changes to weak violet, bright yellow, weak yellow, and gray, respectively [50]. All the samples show excellent visible light absorption as shown in DRS spectra, together with high deNO_x activity under visible light irradiation (Fig. 5.11). Without light irradiation, as shown by the P25 titania sample, the concentration of NO_x returns back to the original value, indicating no chemical reaction after turning off light. Also, the chemical bonding between titanium and nitrogen can be confirmed by the XPS analysis, indicating the solvothermal process is effective for the anion doping under mild conditions in solution.

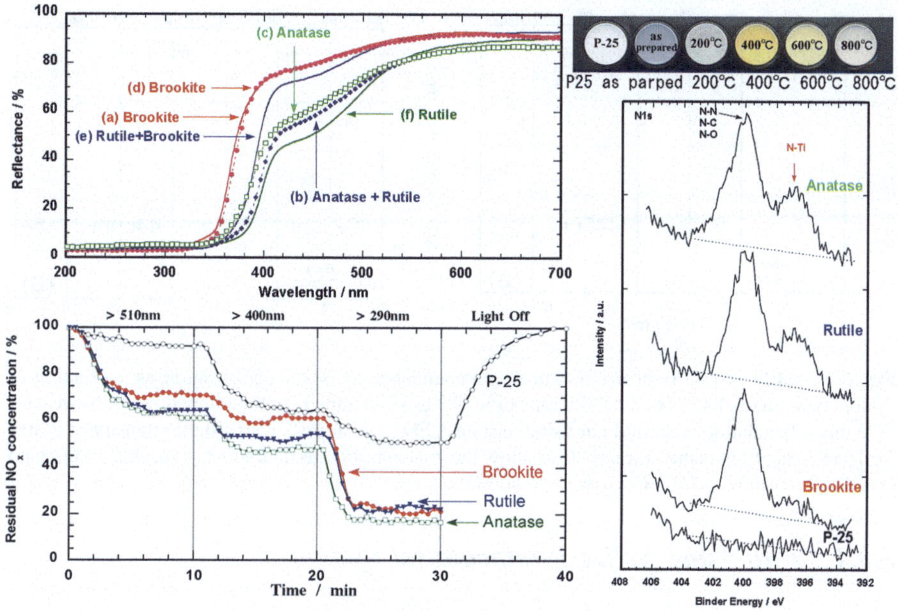

Fig. 5.11 DRS, photocatalytic deNO$_x$ activity, color images, and XPS spectra of anatase, rutile, and brookite phase $TiO_{2-x}N_y$ powders prepared by solvothermal reaction at 190 °C for 2 h in $TiCl_3$–HMT system. Reproduced from Ref. [23, 50] with permission from The RSC & Elsevier

5.3.3 LED Light-Induced Photocatalyst

Solvothermal-synthesized nitrogen-doped photocatalyst particles possess high crystallinity, small particle size, and high specific surface area and show excellent weak light-induced photocatalytic activity. Figure 5.12 shows the irradiation time dependence of the concentration of NO_x in the presence of photocatalysts irradiated by various wavelength of LED lamps [51]. Usually strong light sources are utilized during the characterization, while in this case, even with the use of very weak 2.5 W of monochromatic LED lamps, high deNO$_x$ photocatalytic activity is realized. Non-doping titania (S-TiO_2) shows excellent UV light-induced activity but very low photocatalytic activity under blue light irradiation, while nitrogen-doped titania TiON loaded with platinum shows excellent UV light and visible light-induced photocatalytic activity under both blue light (445 nm) and green light (530 nm) irradiation. It is notable that high deNO$_x$ abilities can be reached even under red light (627 nm) irradiation.

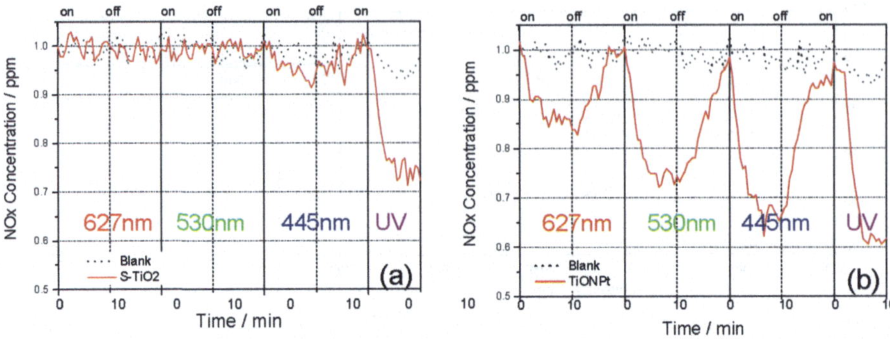

Fig. 5.12 LED visible light irradiation time dependence of NO_x concentration as a function of sample type: (**a**) S-TiO_2, (**b**) TiONPt. One ppm NO gas is pumped continuously through the reactor (373 cm^3). Irradiation is carried out using various LED light sources with a light intensity of 2 mW for 10 min; the light is then turned off to allow the concentration to return to 1 ppm. (Reprinted with permission from Ref. [51]. Copyright 2008 ACS)

5.3.4 Full-Time Active Photocatalyst System

Generally, photocatalyst does not work without light irradiation in night. In order to effectively utilize the nighttime for atmosphere purification, a full-time active photocatalytic system is suggested [10, 52, 53]. The coupling of photocatalysts with a proper long afterglow phosphor such as $CaAl_2O_4$:(Eu, Nd) is expected to prolong the photocatalytic activity even after turning off the light, by using the fluorescence materials as the light source [52].

Figure 5.13 shows the schematic illustration of the concept of the multifunctional persistent full-time reactive photocatalytic system, together with the deNO_x behaviors of $CaAl_2O_4$:(Eu, Nd)/$TiO_{2-x}N_y$, $TiO_{2-x}N_y$, and $CaAl_2O_4$:(Eu, Nd)/undoped TiO_2 (P25) under UV light irradiation followed by turning off the light. This photocatalytic system not only works under light irradiation but also in night without light irradiation. Anion-doped titania shows absorption of visible light up to 700 nm, showing a nice overlap between the diffuse reflectance spectra of mixed anion type photocatalysts and the emission spectrum of long afterglow phosphor $CaAl_2O_4$:(Eu, Nd). After turning off the light irradiation, the persistent deNO_x activity can be continuously confirmed for several hours, which is dependent on the various long afterglow phosphor materials. Besides $CaAl_2O_4$:(Eu, Nd) (emission peak at 440 nm), other long afterglow materials such as $Sr_2Mg(Si_2O_7)$:Eu (peak: 465 nm), $Sr_4Al_{14}O_{25}$:(Eu, Dy) (peak: 490 nm), and $SrAl_2O_4$:(Eu, Dy) (peak: 520 nm) can also be utilized in the proposed system. For not only NO_x and VOC (such as CH_3CHO) but also some dyes in solution (such as MO and RhB dyes solutions), the persistent photocatalytic purification effect with light irradiation can be confirmed [52].

The inset of the figure shows the apparent quantum efficiencies (QE_λ) of the $CaAl_2O_4$:(Eu, Nd)/$TiO_{2-x}N_y$ composite consisting of 40% $TiO_{2-x}N_y$ under UV light irradiation together with that after turning off light, respectively. The quantum

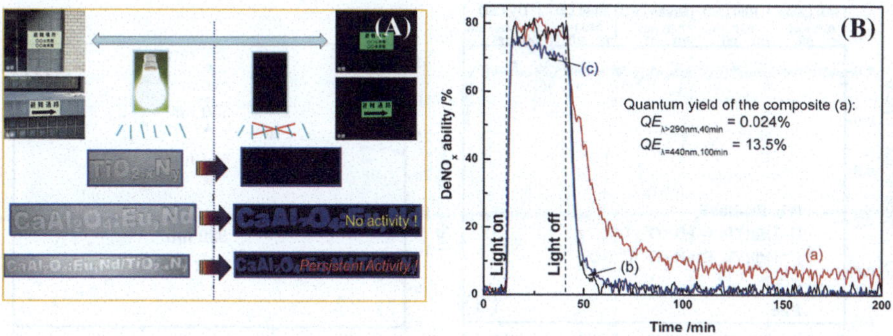

Fig. 5.13 (**A**) Schematic illustration of full-time reactive photocatalytic system; (**B**) photocatalytic deNO$_x$ abilities of (*a*) $CaAl_2O_4$:(Eu, Nd)/$TiO_{2-x}N_y$ composite, (*b*) $CaAl_2O_4$:(Eu, Nd)/TiO_2(P25), and (*c*) pure $TiO_{2-x}N_y$ under UV light irradiation for 30 min followed by turning off light, while NO gas is continuously flowed. The inset shows the quantum yield of the sample (*a*) $CaAl_2O_4$:(Eu, Nd)/$TiO_{2-x}N_y$ composite at 40 min (under UV light with the wavelength of 290 nm for 30 min) and 100 min (in dark after turning off light for 1 h). (Reproduced from Ref. [10, 53] with permission from Ceramic Soc. Jpn. & Elsevier)

yield of the composite at 40 min (under UV light irradiation of 290 nm for 30 min) and 100 min (in dark after turning off light for 1 h, under 440 nm long afterglow luminesce light irradiation) shows the quantum yield of 0.024% and 13.5%, respectively, indicating the high quantum efficiency of the composite. The apparent quantum efficiency (QE_λ) can be calculated using the following equation [53] (Eq. 5.15), which takes into account the deNO$_x$ abilities and average absorption ratios of various light wavelengths ($\lambda > 290$ nm, $\lambda = 440$ nm).

$$QE_\lambda(\%) = (F_{NO} \cdot \alpha_\lambda / P_\lambda \cdot S \cdot A_\lambda) \times 100 \tag{5.15}$$

where F_{NO} (μmol s^{-1}) is the flow quantity of NO molecules in the reaction gas (ml min^{-1}), $\alpha_\lambda(\%)$ the deNO$_x$ ability of the photocatalyst, P_λ (μmol m^{-2} s^{-1}) the light intensity (photon numbers) on the surface of the sample, $S(m^2)$ the surface area of the sample, and $A_\lambda(\%)$ the average light absorption ratio of the sample at different light wavelengths.

5.3.5 *Full-Spectra Active Photocatalyst*

Figure 5.14 shows the deNO$_x$ photocatalytic activity under long wavelength light irradiation, together with their composition dependence. The weak visible light-induced anion-doped titania (such as C-TiO_2) photocatalyst is combined with up-conversion phosphor ((Yb, Er)-$NaYF_4$). The composite can effectively use not only the UV and visible light but also the near-infrared (NIR) light for photocatalytic applications. For pure C-TiO_2, although excellent deNO$_x$ activity is observed under

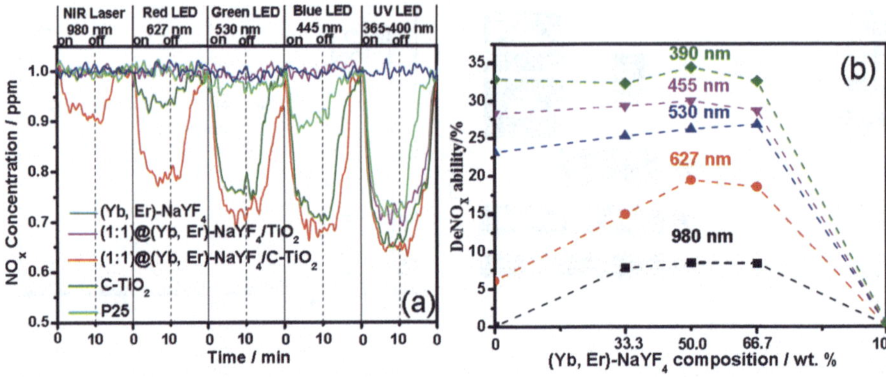

Fig. 5.14 (**a**), (**b**) Photocatalytic deNO$_x$ activities of C-TiO$_2$/(Yb, Er)-NaYF$_4$ under irradiation of visible light LED and NIR laser light. (Reproduced from Ref. [54] with open access to the Nature publishing group)

various visible LED light irradiation until 627 nm, no NIR light-induced activity can be observed, because C-TiO$_2$ cannot be excited by 980 nm NIR light directly. There is no deNO$_x$ activity for (Yb, Er)-NaYF$_4$ alone under irradiation of various wavelength light, because (Yb, Er)-NaYF$_4$ is not a photocatalyst. However, after combination of C-TiO$_2$ and (Yb, Er)-NaYF$_4$ up-conversion phosphor together, continuous deNO$_x$ photocatalytic activity is realized under NIR diode laser irradiation [54, 55].

5.4 Physical and Chemical Properties of Morphologically Controllable Zinc Oxide Ceramics

Zinc oxide ceramics possess many unique physical and chemical properties. It is used not only as a photocatalyst but also as solar cells, sensors, pigments, UV shielding materials, etc. The morphological control of ZnO is an interesting topic. Various morphologies such as nanobelts, nanowires, nanorings, nanocombs, nanosprings, nanopropellers, tetrapods, and tetraleg structures are successfully prepared by so-called "solid-state thermal sublimation" processes [56]. Solution method has also successfully synthesized some unique morphologies and thin films of ZnO with comparatively large scale [57].

Figure 5.15 shows the superhydrophobicity and superhydrophilicity of the ZnO films with different superstructures. It is observed that ZnO films with different superstructures possess different hydrophilicity (i.e., contact angles for a water drop). The ZnO film with nanorod structures (left figure) shows a contact angle of 165° and possesses superhydrophobicity. On the other hand, the ZnO film with nanoscrew structures (right figure) possesses superhydrophilicity, i.e., super wetting properties for water drops. When a drop of water (ϕ2.5 mm) falls on the surface of the film, it rapidly disperses to almost the full area of the substrate (contact angle = 0°).

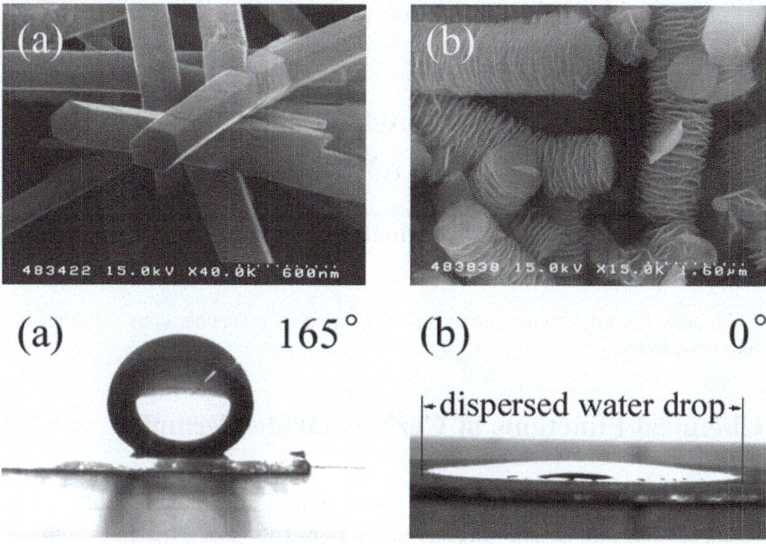

Fig. 5.15 Superhydrophobic and superhydrophilic surface of ZnO films caused by their different superstructures. (**a**) Thin film with nanorod structure, (**b**) thin film with nanoscrew structure

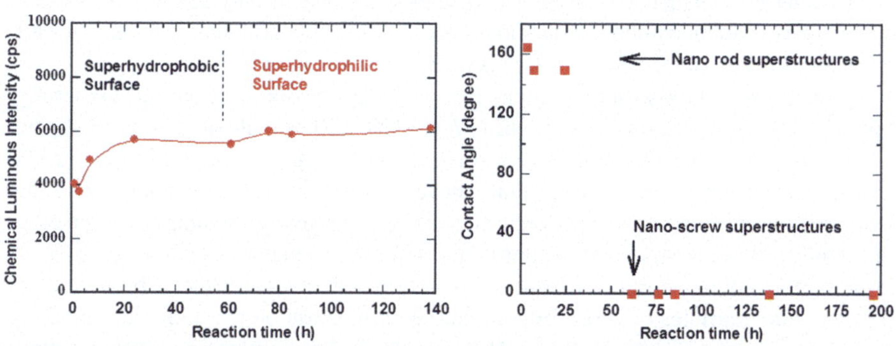

Fig. 5.16 (Left) Chemical luminous intensity and (Right) contact angle of water drops of ZnO films with various morphologies prepared by heating Zn^{2+}-HMT aqueous solution at 95 °C for different reaction times. (Ref. [58] with permission from the Trans Tech Publications)

Figure 5.16 shows the contact angle and chemical luminous intensity of the ZnO thin films with different morphologies. It is found that the film prepared until 24 h in solution possesses nanorods superstructure and shows superhydrophobicity, while that consisting of nonoscrews prepared by aging for more than 36 h shows superhydrophilicity. Interestingly, the chemical luminescence intensity measurement (see right figure) also agrees with the change of the superstructure, indicating that the large surface area of the ZnO film is preferred to cause higher chemical luminescence intensity, in other words, more amount of oxygen related radicals, and of course leads to potential higher chemical reactivity [58].

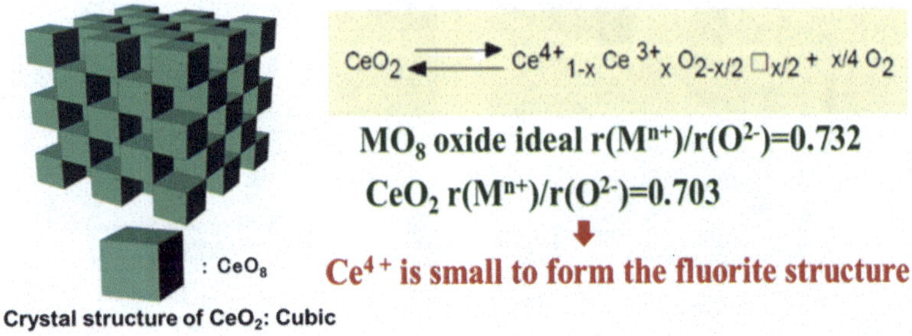

Fig. 5.17 Fluoride crystal structure of cerium oxide and its reversible conversion reaction to a nonstoichiometric oxide

5.5 Chemical Functions of Cerium Oxide Ceramics

Cerium oxide is an oxide of the rare earth metal cerium. It is one of the most important commercial products and an intermediate in the purification of other rare earth elements from the ores. The principal traditional application of ceria is for polishing reagents, especially chemical-mechanical planarization of glass. The distinctive property of this material is its reversible conversion to a nonstoichiometric oxide and also leads to the applications as an oxidation catalyst. The oxidation catalytic activity of CeO_2 is related to the oxygen evolution equilibrium as shown in Fig. 5.17. The ideal ionic size ratio of MO_8 eight coordination oxide $r(M^{n+})/r(O^{2-})$ is 0.732.

Cerium oxide is also an excellent inorganic UV absorber material. Generally, organic UV absorbers show excellent UV-B (290–320 nm) absorption property but with modest UV-A (320–400 nm) absorbing ability. As one of the inorganic UV absorbers, cerium oxide is safer than those of organic ones, by considering the percutaneous absorption. Also, cerium oxide shows many advantages than other inorganic UV absorbers such as titania and zinc oxide fine particles which possess high photocatalytic activities and pose a safety problem when used at high concentrations. Fine particles of ceria show potential application on UV radiation blocking material in personal care products. Because of its lower refractive index, cerium oxide has characteristics which are ideal in a broad spectrum as inorganic sunscreen. Not only high transparency to visible light region but also excellent ultraviolet radiation absorption properties can be realized.

5.5.1 UV Shielding Property and Depression of Its Oxidation Catalytic Activity for Cosmetics

Because of comparatively small ion size of Ce^{4+}, the $r(Ce^{4+})/r(O^{2-})$ is 0.702, which is not big enough to stabilize the fluorite structure. A part of Ce^{4+} needs to be reduced to Ce^{3+} with large ion size, releases the oxygen molecules accompanied by the

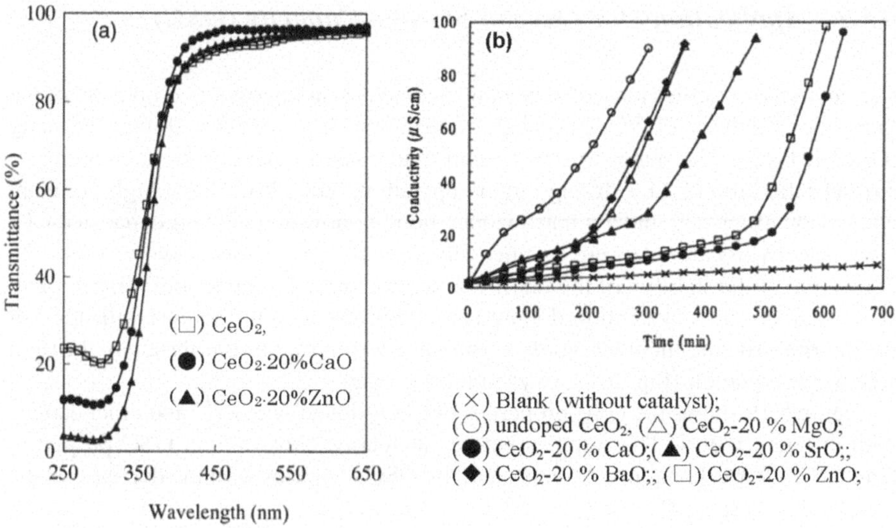

Fig. 5.18 (**a**) UV–Vis transmittance spectra of cation doped CeO_2 (doping amount in mol%) thin films and (**b**) catalytic activity evaluated at 120 °C using the Rancimat system. (Reproduced from Ref. [59] with permission from the Elsevier)

formation of oxygen vacancies, and shows high oxidation catalytic activity. On the other hand, in the case of cosmetic applications, low oxidation activity on the surface of skin is required. In order to establish a more stable eight coordination in the fluorite structure, some large ionic size cation doping is effective to stabilize the fluorite structure, in other words, resulting in a shift of the equilibrium to the left side of the equation (CeO_2 side) [59]. As shown in the right part of Fig. 5.18, the catalytic activity at 120 °C using the Rancimat system (an instrument for measuring the oxidation stability of natural oil) [60] indicates that the catalytic oxidation activity of CeO_2 can be effectively depressed by CaO and ZnO doping. Also the low photocatalytic activity can be realized by the large ionic size of CaO or ZnO doping (although the results are not shown here). In addition, 20 mol% CaO-doped ceria nanoparticles show an excellent UV absorption capacity and high transparency in visible light region compared with that of undoped ceria (left of Fig. 5.18). The high conductivity in the Rancimat test means the high oxidation catalytic activity of the samples. The potential cosmetic material candidates should possess low conductivity and long induction time during the test (Right of Fig. 5.18).

The CaO doping results in obvious decrement of particle size and therefore leads to the formation of a dense film to increase the UV light absorption and decrease the scattering of the light, which is strongly related to the transmittance of visible light. In the case of ZnO doping, the film possesses an excellent UV absorption capacity but inadequate transparency in the visible light regions. Nanoparticles of CaO-doped CeO_2 can be safely used to provide a high SPF (sun protection factor) while maintaining a natural appearance when they are applied to human skin.

5.5.2 Applications for Oxygen Storage Capacity (OSC)

Cerium oxide ceramics have also been recognized as an important component of the three-way catalysts (TWCs) because of their excellent oxygen storage capacity (OSC) [61–64]. The major role of cerium oxide-based materials is maintaining an air/fuel ratio around 14.7 during engine operation, and TWCs have high-pollutant conversion efficiency under a narrow operation window and convert exhausted CO, NO_x, and hydrocarbon to environmentally friendly gases, respectively. From the view point of the stability of the fluorite structure, instead of large ionic size doping, some smaller ionic size cation doping can effectively lead to the unstabilization of the fluorite structure, in other words, resulting in a shift of the equilibrium to the right side of the equation (Fig. 5.17, oxygen release side).

Figure 5.19 shows the OSC properties of CeO_2 doped with Zr^{4+} and other cations with different radius. The samples before calcination show higher OSC properties than those after calcination. As expected, the OSC increases with the decrease of the ionic radius, especially for the Zr^{4+}-Co^{2+} and Zr^{4+}-Sn^{4+} co-doping($Ce_{0.5}Zr_{0.5}O_2$, $Ce_{0.5}Zr_{0.4}M_{0.1}O_2$(M= Co, Sn, Fe)). On the other hand, the OSC properties of CeO_2, $Ce_{0.5}Zr_{0.4}Ti_{0.1}O_2$, $Ce_{0.5}Zr_{0.4}Fe_{0.05}$ $Nb_{0.05}O_2$, and $Ce_{0.5}Zr_{0.4}Bi_{0.05}$ $Nb_{0.05}O_2$ decrease greatly after calcination, indicating that no stable solid solution can be obtained when excessively small cations are doped. Co and Sn doping or loading on thermally

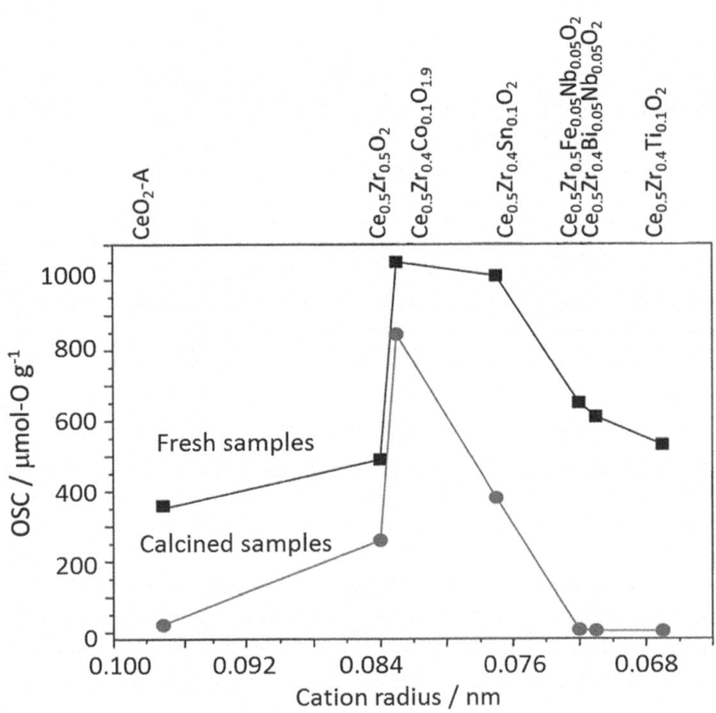

Fig. 5.19 Oxygen storage capacities of various cerium oxides as a function of doped cation radius. (Reproduced from Ref. [63] with permission from the RSC)

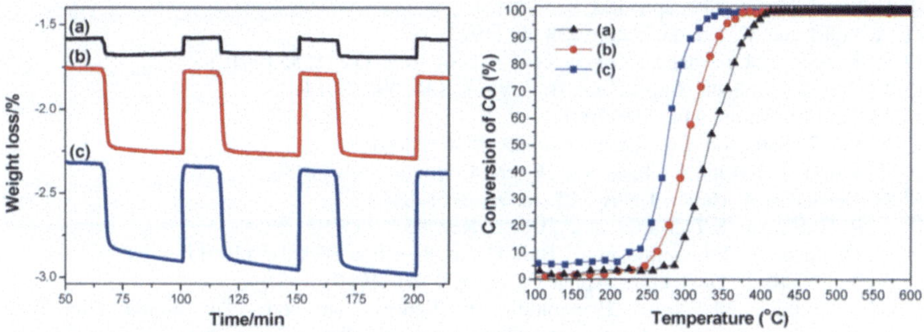

Fig. 5.20 Oxygen release/storage properties (TG profiles) and CO oxidation activity curves of (a) $CeO_2/\gamma\text{-}Al_2O_3$, (b) $Ce_{0.5}Zr_{0.5}O_2/\gamma\text{-}Al_2O_3$, and (c) $Ce_{0.5}Zr_{0.4}Sn_{0.1}O_2/\gamma\text{-}Al_2O_3$ at 600°C. (Reproduced from Ref. [64] with permission from the RSC)

stable alumina matrix results in excellent oxygen storage capacity and shows great potential in CO oxidation activity (Fig. 5.20).

5.6 Remarks and Outlook

Physical and chemical properties of ceramics are greatly affected by their particle size and morphologies. The functions of ceramics can be designed or controlled by environmentally friendly physical and chemical synthesis processes, such as solvothermal synthesis and mechanochemical doping process, etc. Some novel functions such as UV/Vis/IR light-induced photocatalysis, functional thin films with hydrophilicity and hydrophobicity, controllable surface chemical activity, and catalytic activities can be realized by controlling the particle size, crystalline plane, morphology, and components design.

References

1. H.B. Callen, *Thermodynamics and an Introduction to Thermostatistics*, 2nd edn. (Wiley, New York, 1985)
2. J.J. Gilman, J. Appl. Phys. **31**, 2208 (1960)
3. S.L. Bifelow, H.M. Trimble, J. Phys. Chem. **31**, 1798 (1927)
4. G. Buckton, A.E. Beezer, Inter. J. Pharmaceutics **82**, R7 (1992)
5. W. Ostwald, Zeitschrift für physikalische Chemie **22**, 289 (1897)
6. P. Mahon., https://iformulate.biz/news-and-views/dissolution-solution-solubility-stable-formulations/ (2014).
7. P. Antoniammal, D. Arivuoli, J. Nanomater. **2012**, 415797 (2012)
8. P. Buffat, J.P. Borel, Physical Review A **13**, 2287 (1976)
9. T. Sato, P. Zhang, et al., Prog. Cryst. Growth, 58, 92 (2012).

10. S. Yin, J. Ceram. Soc. Jpn. **123**, 823 (2015)
11. A. Fujishima, K. Honda, Nature **238**, 37 (1972)
12. A. Kubacka, M. Fernandez-Garcia, et al., Chem. Rev. **112**, 1555 (2012)
13. I. Altın, M. Sökmen, Appl, Catal. B: Environ. **144**, 694 (2014)
14. M. Grätzel, Nature **414**, 338 (2001)
15. S. Yin, T. Sato, Ind. Eng. Chem. Res. **39**, 4526 (2000)
16. T. Kawai, T. Sakata, J. Chem. Soc., Chem. Commun., 694 (1980).
17. H. Gerischer, A. Heller, J. Phys. Chem. **95**, 5261 (1991)
18. M.R. Hoffmann, S.T. Martin, et al., Chem. Rev. **95**, 69 (1995)
19. K. Okamoto, Y. Yamamoto, et al., Bull. Chem. Soc. Jpn. **58**, 2015 (1991)
20. V. Auguliaro, E. Davi, et al., Applied Catal. **65**, 101 (1990)
21. M. Takeuchi, M. Anpo, Preparation of Highly Transparent TiO2-based Thin Film Photocatalysts by an Ion Engineering Method: Ionized Cluster Beam Deposition, in *Environmentally Benign Photocatalysts*, ed. by M. Anpo, P. V. Kamat, (Springer, New York, 2010), pp. 133–152
22. S. Yin, Y. Inoue, et al., J. Mater. Res. **13**, 844 (1998)
23. S. Yin, Y. Aita, et al., J. Mater. Chem. **15**, 674 (2005)
24. S. Yin, H. Yamaki, et al., J. Mater. Chem. **13**, 2996 (2003)
25. P. Chowdhury, H. Gomaa, et al., ACS Symposium Series, 1124, Sustainable Nanotechnology and the Environment: Advances and Achievements, Najm Shamim, Virender K. Sharma Chapter 13: Dye-Sensitized Photocatalyst: A Breakthrough in Green Energy and Environmental Detoxification, American Chemical Society: Washington, DC, pp. 231 (2013).
26. L. Wang, J. Shang, et al., Scientific Reports **4**, 7384 (2014)
27. E. Borgarello, J. Kiwi, et al., J. Am. Chem. Soc. **103**, 6324 (1981)
28. R. Asahi, T. Morikawa, et al., Chem. Rev. **114**, 9824 (2014)
29. T. Morikawa, R. Asahi, et al., Jpn. J. Appl. Phys. **40**, 6A (2001)
30. S. Yin, T. Sato, J. Photochem. Photobiol. A: Chem. **169**, 89 (2005)
31. M. Murdoch, G.I.N. Waterhouse, et al., Nature Chem. **3**, 489 (2011)
32. J. Wu, J. Lin, et al., J. Mater. Chem. **11**, 3343 (2001)
33. S. Yin, D. Maeda, et al., Solid State Ionics **151**, 377 (2002)
34. K.R. Gopidas, M. Bohorquez, et al., J. Phys. Chem. **94**, 6435 (1990)
35. K. Maeda, ACS Catal. **3**, 1486 (2013)
36. H. Katsumata, Y. Tachi, et al., RSC Adv. **4**, 21405 (2014)
37. Z. Zhao, J. Fan, et al., Sci. China Technol. Sci. **60**, 1 (2017)
38. T. Takata, C. Pan, et al., Sci. Technol. Adv. Mater. **16**, 033506 (2015)
39. H. Kato, K. Ueda, et al., J. Mater. Chem. A **3**, 11824 (2015)
40. I. Justicia, P. Ordejon, et al., Adv. Mater. **14**, 1399 (2002)
41. T. Ihara, M. Miyoshi, et al., J. Mater. Sci. **36**, 4201 (2002)
42. H. Irie, Y. Watanabe, et al., Chem. Lett. **32**, 772 (2003)
43. T. Umebayashi, T. Yamaki, et al., Appl. Phys. Lett. **81**, 454 (2003)
44. T. Ohno, M. Akiyoshi, et al., Appl. Catal. A: General **265**, 115 (2004)
45. T. Ikoma, Q. Zhang, et al., Bull. Chem. Soc. Jpn. **74**, 2303 (2001)
46. J. Wang, S. Yin, et al., J. Mater. Chem. **13**, 2348 (2003)
47. S. Yin, Q. Zhang, et al., Chapter 13: Synthesis of Titanium Dioxide-Based, Visible-Light Induced Photocatalysts by Mechanochemical Doping, in *High Energy Ball Milling: Mechanochemical Processing of Nanopowders*, ed. by M. Sopicka-Lizer, (Woodhead Publishing, Oxford, 2010), p. 304
48. S. Yin, H. Yamaki, et al., Solid State Sciences **7**, 1479 (2006)
49. K. Byrappa, M. Yoshimura, "Handbook of Hydrothermal Technology", 2nd, William Andrew Publishing, Oxford (2013).
50. S. Yin, Y. Aita, et al., J. Eur. Ceram. Soc. **26**, 2735 (2006)
51. S. Yin, B. Liu, et al., J. Phys. Chem. C **112**, 12425 (2008)
52. H.H. Li, S. Yin, et al., Funct. Mater. Lett. **6**, 1330005 (2013)

53. H.H. Li, S. Yin, et al., Appl. Catal. B: Environ. **106**, 586 (2011)
54. X. Wu, S. Yin, et al., Scientific Reports **3**, 2918 (2013)
55. X. Wu, S. Yin, et al., Appl. Catal. B: Envoron. **156–157**, 257 (2014)
56. X. Wang, J. Song, et al., J. Mater. Chem. **17**, 711 (2007)
57. S. Yin, T. Sato, J. Mater. Chem. **15**, 4584 (2005)
58. S. Yin, R. Li, T. Sato, Adv. Sci. Technol. **45**(4), 679–684 (2006)
59. S. Yabe, M. Yamashita, et al., Int. J. Inorg. Mater. **3**, 1003 (2001)
60. Rancimat, in: https://www.metrohm.com/en-us/products-overview/stability-measurement/
61. M. Ozawa, M. Kimura, et al., J. Alloys Compd. **193**, 73 (1993)
62. J. Wu, H. Chen, et al., J. Catal. Rev. Sci. Eng. **57**, 79 (2015)
63. Q. Dong, S. Yin, et al., RSC Advances **2**, 12770 (2012)
64. Q. Dong, S. Yin, et al., Catal. Sci. Technol. **2**, 2521 (2012)

Chapter 6
Biological Functions of Ceramics

Masakazu Kawashita

Abstract Technologies applied to characterizing ceramics for biomedical applications have essential roles in the development of artificial bones and teeth. By recovering biological function hampered by aging and disease and maintaining a high quality of life for patients, healthcare is of increasing significance in an aged society such as Japan. At present, healthcare includes not only technologies that sustain life but also those that can extend healthy lives. In this chapter, ceramics useful for recovering biological function are introduced.

Keywords Artificial bone · Bioactive ceramics · Apatite · Simulated body fluid

6.1 Requirements of Artificial Materials for Bone Repair

The human body loses function when it is damaged by disease, injury, or aging. Artificial materials that are used in contact with the body surface, its inner tissue, or body fluids in order to recover or support the body's function are called biomaterials [1]. When the bone is damaged, for example, motor function is lost, and the patient's quality of life (QOL) is significantly diminished. In such cases, artificial bone that substitutes the function of natural bone is used clinically. As bone substitutes, ceramic materials have an important role in clinical practice, especially orthopedics.

There are several requirements for biomaterials to recover bone function, as listed below:

A. Biological properties

(i) No toxicity, no tissue inflammation, and no carcinogenicity
(ii) Affinity to surrounding soft tissue
(iii) Strong chemical bonding to surrounding bone

M. Kawashita (✉)
Department of Inorganic Biomaterials, Institute of Biomaterials and Bioengineering, Tokyo Medical and Dental University, Tokyo, Japan
e-mail: kawashita.bcr@tmd.ac.jp

J. Hojo (ed.), *Materials Chemistry of Ceramics*,
https://doi.org/10.1007/978-981-13-9935-0_6

B. Mechanical properties

(i) Long-term mechanical properties analogous to or higher than bone
(ii) Elastic modulus similar to bone
(iii) Low friction coefficient and low rate of friction at joint surface

C. Requirements for use in clinical practice

(i) Easy molding and working
(ii) No deterioration by sterilization such as autoclave
(iii) Easy handling in operation room

Regarding the biocompatibility of biomaterials for bone repair, both mechanical and biological compatibilities must be considered. These compatibilities should be evaluated in terms of the implantation site of the biomaterial or its substitution for body tissues. Actually, the required mechanical properties of biomaterials for cortical bone are very different from those for cancellous bone. For cortical bone, the compressive strength, bending strength, Young's modulus, and fracture toughness (K_{IC}) are 100–230 MPa, 50–150 MPa, 7–30 GPa, and 2–12 MPa $m^{1/2}$, respectively, whereas the compressive strength, bending strength, and Young's modulus for cancellous bone are 2–12 MPa, 10–20 MPa, and 0.05–0.5 GPa, respectively. In addition to evaluating biocompatibility, both short- and long-term evaluation in the body is needed.

Ceramics used clinically can be categorized into three materials based on their biological response: bioinert, bioactive, and bioabsorbable (biodegradable). Table 6.1 shows ceramics for bone and joint repair with some representative examples. Bioinert materials form collagen layers on their surfaces, but the collagen layer is extremely thin. Hence, these materials appear to contact the bone under an optical microscope. However, bioinert materials do not bond to bone directly; their fixation at the implant site is achieved by mechanical interlocking. Bioinert materials are mainly used for femoral components in artificial joints. By comparison, bioactive materials form strong bonds to bone without a foreign body response. They are used as artificial iliac bones and artificial vertebra, but their use is limited to unloaded

Table 6.1 Ceramics for bone and joint repair with some representative examples

Bioinert material	Sintered alumina (Al_2O_3)
	Sintered zirconia (ZrO_2)
Bioactive material	Bioglass® (Na_2O-CaO-SiO_2-P_2O_5 glass)
	Ceravital® (Na_2O-K_2O-MgO-CaO-SiO_2-P_2O_5 glass-ceramics, precipitated crystalline phase: oxyapatite)
	Cerabone® A-W (Glass-ceramic A-W) (MgO-CaO-SiO_2-P_2O_5-CaF_2 glass-ceramics, precipitated crystalline phase: oxyfluoroapatite)
	Sintered hydroxyapatite ($Ca_{10}(PO_4)_6(OH)_2$)
Bioabsorbable material	Sintered β-tricalcium phosphate ($3CaO{\cdot}P_2O_5$)
	Limestone ($CaCO_3$)

implantation sites because their mechanical properties differ considerably from living bone. Bioabsorbable materials decompose and are gradually absorbed in the body and replaced with living bone, especially when they are implanted into bone defects. However, the mechanical properties of conventional bioabsorbable ceramics remain vastly different from those of bone, and new technologies that can control the absorption rate are desirable.

Among several characteristics of ceramics, their bone-bonding ability (i.e., bioactivity) is especially unique. Bioactive ceramics have been investigated extensively as biomaterials showing specific biological functions to bone tissue, and several attempts have been made to apply this function to novel bone-repairing materials. Understanding the bone-bonding mechanism of bioactive ceramics has contributed to the development of regenerative medicine as well as orthopedics and dentistry.

6.2 Bone-Bonding Ceramics

When artificial materials are implanted into bone defects, they are generally encapsulated by collagen fibers to be isolated from the surrounding bone. This strategy is our body's common response to protect itself from foreign materials. Because of this encapsulation, however, artificial materials cannot become securely fixed to the surrounding bone. In the early 1970s, Hench et al. discovered that some glasses in the Na_2O-CaO-SiO_2-P_2O_5 system could bond to living bone tightly without the formation of fibrous tissues [2]. The glasses were named Bioglass®, and their typical composition (45S5) in mol% was 46.1 SiO_2, 24.4 Na_2O, 26.9 CaO, and 2.6 P_2O_5. Since then, various types of bone-bonding materials have been developed and used clinically. Bone-bonding ceramics such as Bioglass® are now called bioactive ceramics. The phenomenon of new bone formation on bioactive ceramics is called osteoconduction. The osteoconductivity of bioactive ceramics can achieve strong bonding between ceramics and bone, providing great potential to the development of stable bone substitute implants.

Glass-ceramics obtained by the crystallization of glass also show bioactivity. A typical example is the glass-ceramic A-W obtained by heat treatment of MgO-CaO-SiO_2-P_2O_5-CaF_2 glass. Sintered hydroxyapatite ($Ca_{10}(PO_4)_6(OH)_2$) is also used clinically as a non-bonding material. Bone is mainly composed of hydroxyapatite (70 mass%) and collagen (30 mass%); hence, it is reasonable to consider that hydroxyapatite can bond to bone. However, even glasses and glass-ceramics whose compositions are entirely different from bone can bond to living bone. Furthermore, Bioglass® and glass-ceramic A-W show higher osteoconductivities than sintered hydroxyapatite [3]. According to previous studies, these glass and glass-ceramics bond to bone through a hydroxyapatite layer. Notably, this apatite layer does not form at the interface between non-bioactive glass and glass-ceramics and bone [4, 5]. As proposed by Kokubo et al. [6], simulated body fluid (SBF) can be utilized to reproduce the formation of an apatite layer on bioactive glass and glass-ceramics. SBF is an acellular protein-free solution and contains inorganic ions with

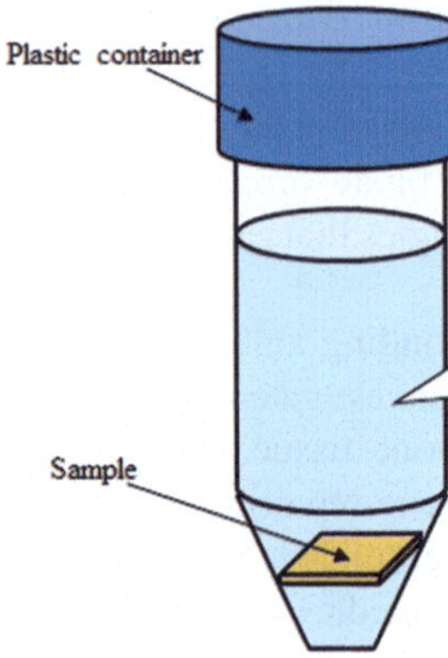

- Temperature: 36.5°C
- pH is controlled at 7.25 or 7.40 by 50 mol·m^{-3} tris(hydroxymethyl)aminomethane and 45 mol·m^{-3} hydrochloric acid (HCl)

Ion	Concentration (mol·m^{-3})	
	Simulated body fluid (SBF)	Human blood plasma
Na^+	142.0	142.0
K^+	5.0	5.0
Mg^{2+}	1.5	1.5
Ca^{2+}	2.5	2.5
Cl^-	147.8	103.0
HCO_3^-	4.2	27.0
HPO_4^{2-}	1.0	1.0
SO_4^{2-}	0.5	0.5

Fig. 6.1 In vitro evaluation of apatite-forming ability of samples using simulated body fluid (SBF) with ion concentrations nearly equal to those of human blood plasma

concentrations nearly equal to those of human blood plasma. Its pH is controlled at 7.25 or 7.40 by tris(hydroxymethyl)aminomethane and hydrochloric acid (HCl), as shown in Fig. 6.1. Consequently, it is believed that the apatite layer is formed on bioactive ceramics via chemical reactions between the bioactive ceramics and body fluid.

According to the analysis of glass-ceramic A-W surfaces by thin-film X-ray diffractometry and Fourier-transform infrared spectrometry, the apatite layer formed on glass-ceramic A-W was composed of carbonate ion-containing hydroxyapatite microcrystals with defective structures and/or small crystallites [7], and it was similar to bone mineral in its composition and structure. Therefore, it is speculated that fibroblasts are difficult to proliferate on the apatite layer, and hence bone can contact the apatite layer directly without fibrous tissues. After contact, strong chemical bonds are formed between the bone and apatite layer in order to decrease the interfacial surface energy. When two rectangular plates of glass-ceramic A-W tied together with silk thread were soaked in SBF or implanted subcutaneously into rats for 1 month, they strongly bonded to each other through the apatite layer and could not be mechanically separated [8, 9]. This suggests that strong chemical bonds can form between the apatite layers in vivo without the contribution of organic compounds such as proteins and cells.

Based on these findings, bioactive materials should be fabricated by designing their surfaces to form apatite with a composition and structure analogous to bone mineral, which is called bone-like apatite, via chemical methods (Fig. 6.2). The apatite-forming ability of materials in the body fluid environment depends on both the composition and structure of the materials. For example, the rate of apatite formation for some bioactive ceramics is increased in the order: sintered

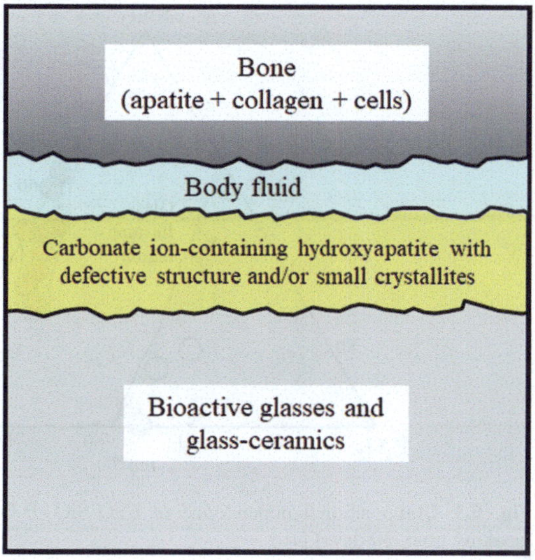

Fig. 6.2 Bone-bonding mechanism of bioactive glasses and glass-ceramics

hydroxyapatite < glass-ceramics A-W < Bioglass®. Sintered hydroxyapatite has a nearly stoichiometric composition ($Ca_{10}(PO_4)_6(OH)_2$), whereas, in bone apatite, PO_4^{3-} ions are partially replaced with CO_3^{2-} and HPO_4^{2-} ions, Ca^{2+} ions are partially replaced with Mg^{2+} and Na^+ ions, and OH^- ions are partially replaced with Cl^- ions. Therefore, even sintered hydroxyapatite forms bone-like apatite on its surface when it is implanted into the body, which remains stable under body fluid conditions. However, the rate of reaction is low because the reaction progresses by ion exchange between hydroxyapatite crystals and body fluid. In contrast, glass-ceramic A-W contains a CaO-SiO_2 glassy phase besides oxyfluoroapatite ($Ca_{10}(PO_4)_6(O, F)_2$) and β-wollastonite ($CaSiO_3$). The CaO-SiO_2 glassy phase plays an important role in the formation of bone-like apatite in the body fluid environment, as described in Sect. 6.3.

6.3 Mechanism of Apatite Formation on the Surface of Bioactive Ceramics

Figure 6.3 shows the compositional dependence of CaO-SiO_2-P_2O_5 glasses on apatite formation in SBF [10]. Interestingly, CaO-SiO_2 glasses without P_2O_5 (typically 50CaO·50SiO_2 (mol%) glass) as well as those containing small amounts of P_2O_5 form apatite on their surfaces, whereas CaO-P_2O_5 glasses without SiO_2 as well as those containing SiO_2 do not. Body fluid contains calcium (Ca^{2+}) and phosphate ions with concentrations exceeding the solubility of hydroxyapatite ($Ca_{10}(PO_4)_6(OH)_2$), even under normal conditions (i.e., our body fluid is supersaturated with respect to hydroxyapatite). When 50CaO·50SiO_2 (mol%) glass is

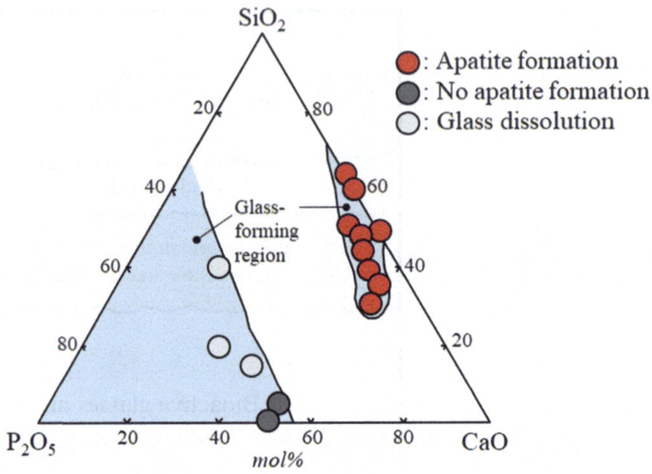

Fig. 6.3 Compositional dependence of CaO-SiO_2-P_2O_5 glasses on apatite formation in SBF (soaking time, 30 days) [10]

exposed to body fluid, Ca^{2+} ions are released from the glass to the surrounding fluid. As a result, the degree of supersaturation of the surrounding fluid with respect to hydroxyapatite is increased. The release of Ca^{2+} ions also forms hydrolyzed silica (SiO_2) gel abundant with Si-OH groups on the glass surface, and the Si-OH groups induce the heterogeneous nucleation of apatite in the body fluid environment [11]. During apatite formation, phosphate ions are supplied from the surrounding body fluid; hence, the $50CaO{\cdot}50SiO_2$ (mol%) glass can form apatite although it contains no P_2O_5. That is, the release of Ca^{2+} ions increases the degree of supersaturation of the surrounding fluid with respect to apatite and forms functional groups such as Si-OH groups, which induce the heterogeneous nucleation of apatite. Once the apatite nuclei are formed, they grow spontaneously to form an apatite layer by consuming the Ca^{2+} and phosphate ions from the surrounding fluid. Consequently, the $50CaO{\cdot}50SiO_2$ (mol%) glass exposed to body fluid forms the gradient surface structure shown in Fig. 6.4. Novel bioactive materials can be developed by control of this aforementioned apatite formation process.

6.4 Bioactive Organic-Inorganic Composite Materials

Conventional bioactive ceramics are not able to replace highly loaded bones, such as the femur and tibia, since they have a lower fracture toughness and a higher Young's modulus than bone. Bone has a well-designed structure where inorganic tiny apatite crystals, 20–40 nm in size, are deposited on three-dimensional collagen fibers. Based on the organic-inorganic composite structure of bone, bioactive organic-inorganic composites with plastic deformability have been developed. Bonfield et al.

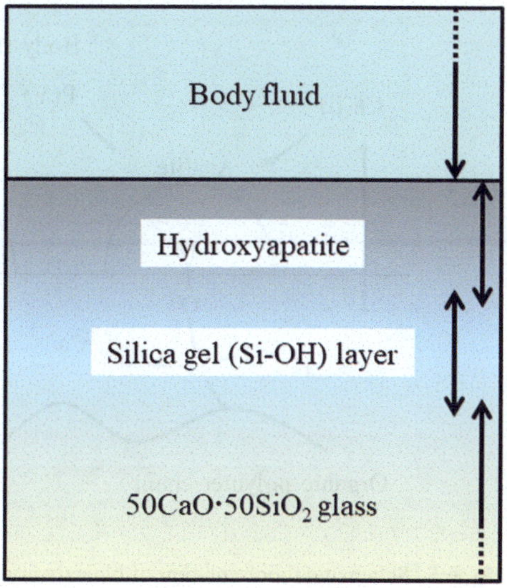

Fig. 6.4 Surface structure of $50CaO \cdot 50SiO_2$ (mol%) glass exposed to body fluid

developed a hydroxyapatite – high-density polyethylene (HDPE) composite in which HDPE is reinforced with hydroxyapatite [12]. The resultant composite demonstrates bioactivity due to the presence of hydroxyapatite and exhibits mechanical properties analogous to cortical bone, resulting in the formation of stable bonds between the composite and bone. Further, the composite can be cut into the desired shape during surgery.

Kikuchi et al. developed a hydroxyapatite/collagen bone-like nanocomposite [13], which is completely incorporated into the bone remodeling process for substitution by new bone. In 2012, a hydroxyapatite/collagen sponge composed of hydroxyapatite (80 wt%) and collagen (20 wt%) received marketing and manufacturing approval in Japan. Octacalcium phosphate (OCP, $Ca_8H_2(PO_4)_6 \cdot 5H_2O$) is a precursor of hydroxyapatite in the mineralization of living bone. Suzuki et al. found that OCP shows better osteoconductivity than hydroxyapatite [14] and developed OCP-based bone substitute materials [15]. The OCP/collagen composite is presently being subjected to clinical trials in Japan.

According to the mechanism of apatite formation on CaO-SiO_2 glass in body fluid described in Sect. 6.3, it is useful to utilize the components releasing Ca^{2+} ions and those that have Si-OH groups to synthesize novel bioactive materials. If organic polymers modify these components, novel biomaterials with osteoconductivity as well as distinct mechanical properties can be obtained. Figure 6.5 shows a schematic representation of organic-inorganic hybrids for bone repair. Such organic-inorganic hybrids can be synthesized by the sol-gel method utilizing the hydrolysis and polycondensation of silicon alkoxide compounds. In this respect, Mackenzie et al. proposed organically modified silicates (ORMOSILS) composed of polydimethylsiloxane (PDMS) and SiO_2 derived from tetraethoxysilane (TEOS) [16]. The ORMOSILS possess chemical and thermal stability due to SiO_2 and flexibility due to PDMS. Osaka et al. developed PDMS-CaO-SiO_2 hybrids showing

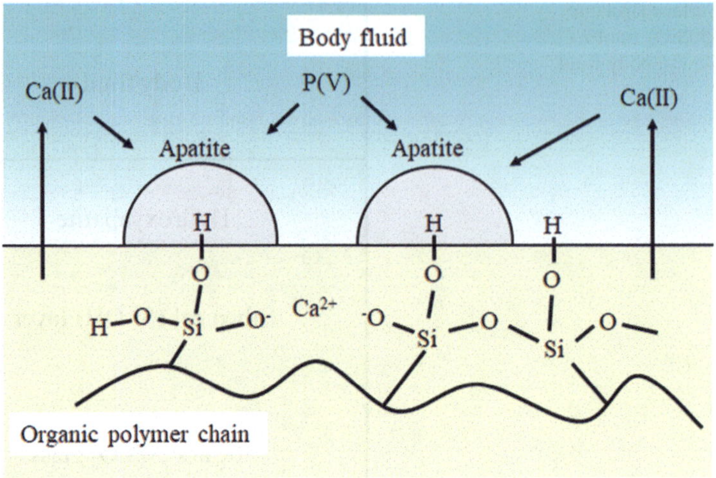

Fig. 6.5 Schematic representation of bioactive organic-inorganic hybrids

an apatite-forming ability in SBF similar to bioactive glass [17]. Regarding the organic component, vinyltrimethoxysilane ($H_2C{=}CHSi(OCH_3)_3$) and 3-methacryloxy propyltrimethoxysilane (MPS; $H_2C{=}CCH_3COO(CH_2)_3Si(OCH_3)_3$) have been considered [18, 19]. MPS and 2-hydroxyethylmethacrylate (HEMA; $H_2C{=}C(CH_3)COOCH_2CH_2OH$) are polymerized, and calcium chloride ($CaCl_2$) is added to the polymer and dried to obtain MPS-HEMA-$CaCl_2$ hybrids, as shown in Fig. 6.6 [20]. If the hybrids are soaked in SBF, those with a MPS:HEMA=0.1:0.9 (molar ratio) composition can form apatite on their surface after 7 days of soaking as shown in Fig. 6.7. These organic-inorganic hybrids have a wide range of biomedical applications because they can be coated on polymeric materials with excellent mechanical properties or blended with conventional polymers. In fact, if MPS and $CaCl_2$ are added to polymethylmethacrylate (PMMA) bone cement, the cement forms apatite in SBF [21]. These organic-inorganic hybrids are believed to be novel biomaterial candidates with the osteoconductivity of bioactive inorganic components and the flexibility of organic components.

6.5 Induction of Bioactivity on Metals by Chemical Treatment

On the basis of the apatite formation mechanism of bioactive ceramics in body fluid, a technology to induce bioactivity directly on titanium and its alloys has been developed. Li et al. found that hydrolyzed titania (TiO_2) and SiO_2 gels formed apatite in SBF [22]. This finding suggests that bioactivity can be induced on titanium when its surface is modified by chemical treatments that enable it to be hydrolyzed to

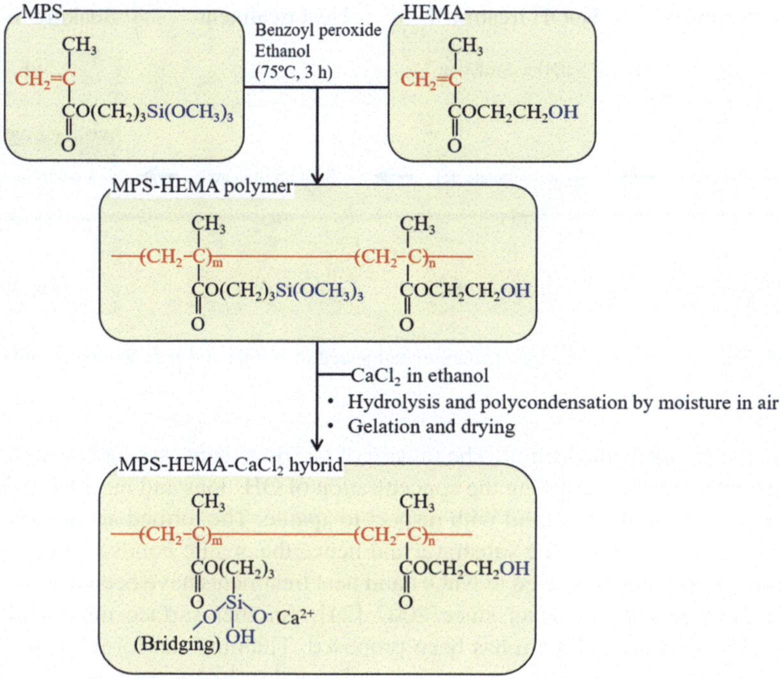

Fig. 6.6 Synthesis of MPS-HEMA-$CaCl_2$ hybrid

Fig. 6.7 Scanning electron micrographs of MPS-HEMA-$CaCl_2$ hybrids soaked in SBF for 7 days (MPS:HEMA=0.1:0.9 molar ratio) [20]

form TiO_2 gel under body fluid conditions. For example, low crystallinity sodium titanate is formed on titanium when titanium is subjected to a 5 kmol m^{-3} sodium hydroxide (NaOH) solution and subsequent heat treatment at 600 °C, as shown in Fig. 6.8. Thus, when treated titanium is subjected to body fluid, sodium (Na^+) ions are released from the titanium surface into the body fluid, and hydronium (H_3O^+) ions are incorporated into the titanium surface to form many Ti-OH groups [23],

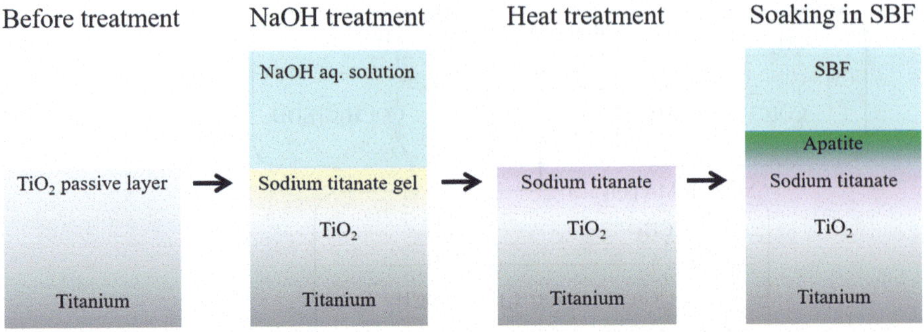

Fig. 6.8 Surface structural change of titanium subjected to NaOH and heat treatment and subsequently soaked in SBF

which induce apatite nucleation. The release of Na^+ ions increases and enhances the apatite nucleation by increasing the concentration of OH^- ions and increasing degree of supersaturation of body fluid with respect to apatite. The formed apatite develops a graded structure toward the substrate, and hence the apatite bonds strongly to the substrates. Hip joints subjected to NaOH and heat treatments have been used in more than 20,000 patients in Japan since 2007 [24]. Another surface modification to induce bioactivity on titanium has been proposed. Titanium can form apatite on its surface in SBF within 14 days when it is subjected to hydrogen peroxide (H_2O_2) solutions containing tantalum chloride ($TaCl_5$) at 60 °C [25]. In this way, treated titanium bonds more strongly to living bone than untreated titanium [26, 27]. As described above, bioactivity can be induced in titanium and its alloys when hydrolyzed TiO_2 gels are formed on their surfaces.

Some attempts to induce bioactivity in other metals and nanocomposites have been conducted. For example, when tantalum is subjected to NaOH and heat treatments, a bioactive surface layer in the form of Ta-OH groups under body fluid environments is formed [28], and the treated tantalum bonds to living bone [29], as described above for bioactive titanium and its alloys. Further, Zr-OH groups with apatite-forming ability can be induced in ceria-stabilized zirconia/alumina nanocomposites when the nanocomposites are subjected to NaOH, HCl, sulfuric acid (H_2SO_4), or phosphoric acid (H_3PO_4) solutions [30].

6.6 Ceramics with Pharmacological Function

Stem cells, growth factors, and scaffolds are three factors that are essential in the construction of living tissues by regenerative medicine or tissue engineering. Bioactive ceramics or bioabsorbable ceramics are utilized as the scaffolds. Additionally, the induction of pharmacological functions to promote bone formation and suppress bone absorption has been attempted for artificial bones. For example, zinc (Zn) ions, which enhance bone formation by osteoblasts and suppress bone absorption by

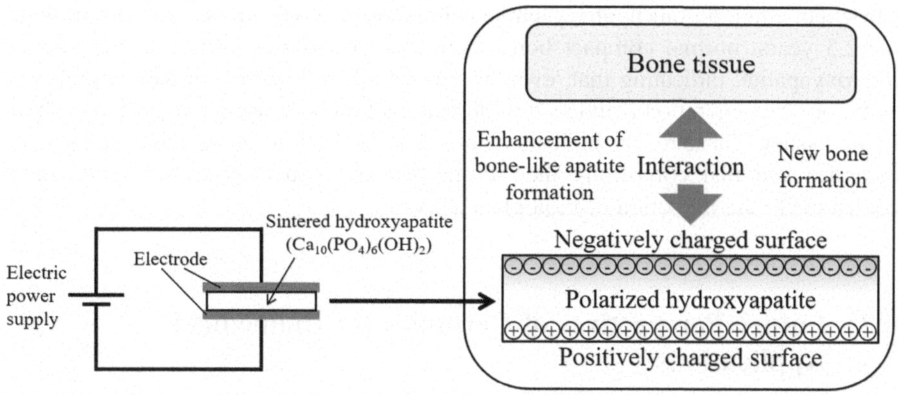

Fig. 6.9 Enhancement of biological activity of sintered hydroxyapatite by polarization

osteoclasts, have been doped to tricalcium phosphate. The effect of the amount of incorporated Zn and the (Zn+Ca)/P ratio on the releasing behavior of Zn has been investigated. When the (Zn+Ca)/P ratio is increased from 1.60 to 1.64, enough Zn is released from the composite to enhance bone formation, even though the Zn content decreases from 0.316 to 0.21 mass% [31]. Further, when Zn-containing β-TCP in 5 wt% is added to calcium phosphate bone cement, proliferation of osteoblastic cells is enhanced [32]. A trace amount of silicon species enhances the proliferation and mineralization of osteogenic cells and mesenchymal stem cells. Kasuga et al. developed electrospun microfiber meshes of a silicon-doped vaterite ($CaCO_3$)/poly(lactic acid) hybrid for bone regeneration [33]. The silicon-doped vaterite has a high solubility and gradually releases silicon species into the body, and the poly(lactic acid) is bioabsorbable. The cotton-like bioabsorbable bone void filler of silicon-doped vaterite/poly-L-lactic acid (PLLA)/β-tricalcium phosphate (β-TCP) composite microfibers was approved by the Food and Drug Administration (FDA) in 2014 and marketed as ReBOSSIS® in the United States in 2015. Various types of novel biomaterials releasing pharmacological trace elements are likely to be developed further in the future.

Yamashita et al. proposed new technology to control the biological function of hydroxyapatite [34]. As shown in Fig. 6.9, sintered hydroxyapatite can be polarized by application of an external dc field at high temperature, with the polarization effect maintained over 6 months after the initial polarization. Therefore, when polarized hydroxyapatite is implanted into the body, a local electric field can be applied to the surrounding living tissues for sustained periods of time. Apatite formation in SBF is enhanced on the negatively charged surface of hydroxyapatite, whereas it is slightly suppressed on the positively charged surface of hydroxyapatite [34]. Additionally, highly oriented bone layers are formed by activated osteoblastic cells on the negatively charged surface of hydroxyapatite [35]. de Groot et al. reported that porous calcium phosphate ceramics such as hydroxyapatite and β-TCP can induce bone formation when they are implanted intramuscularly for long periods of time [36]. In particular, when porous hydroxyapatite is implanted into the dorsal muscles of dogs for 2.5 years, normal compact bone with bone marrow is formed in the porous hydroxyapatite, indicating that, even at the site where bone is not formed intrinsically, bone regeneration is induced by the interaction between ceramics and cells or living tissues. These results indicate that bone regeneration can be achieved only by ceramics and that controlling their properties and structure would significantly contribute to the induction of biological activity.

6.7 Future Perspectives of Ceramics for Biomedical Applications

The design of ceramic-based materials to repair biological functions is described in this chapter from the perspective of bioactive ceramics. The osteoconductivity of bioactive ceramics is effective not only for repairing function of hard tissues but also for extracellular matrix or scaffolds in regenerating skin, nerves, bone, and blood vessels. In addition to ceramics to repair bone, select ceramics for cancer treatments and those with antibacterial activity also have been proposed [37], though discussion of such materials has been omitted from this chapter due to space limitations. In the future, the interactions between materials and living tissues will be revealed more fully, aiding the development of functional ceramic-based biomaterials for improved quality of life.

References

1. D.F. Williams, *The Williams Dictionary of Biomaterials* (Liverpool University Press, Liverpool, 1999), p. 42
2. L.L. Hench, R.J. Splinter, W.C. Allen, T.K. Greenlee, J. Biomed. Mater. Res. Symp. **2**, 117 (1971)
3. K. Ono, T. Yamamuro, T. Nakamura, T. Kokubo, Biomaterials **11**, 265 (1990)
4. C. Ohtsuki, H. Kushitani, T. Kokubo, S. Kotani, T. Yamamuro, J. Biomed. Mater. Res. **25**, 1363 (1991)
5. M. Neo, S. Kotani, T. Nakamura, T. Yamamuro, C. Ohtsuki, T. Kokubo, Y. Bando, J. Biomed. Mater. Res. **26**, 1419 (1992)
6. T. Kokubo, H. Takadama, Biomaterials **27**, 2907 (2006)
7. T. Kokubo, S. Ito, Z. Huang, T. Hayashi, S. Sakka, T. Kitsugi, T. Yamamuro, J. Biomed. Mater. Res. **24**, 331 (1990)
8. T. Kokubo, T. Hayashi, S. Sakka, T. Kitsugi, T. Yamamuro, J. Ceram. Soc. Japan (Yogyo-Kyokai-Shi) **95**, 785 (1987)
9. Y. Fujita, T. Yamamuro, T. Nakamura, T. Kitsugi, S. Kotani, C. Ohtsuki, T. Kokubo, J. Biomed. Mater. Res. **26**, 1311 (1992)
10. C. Ohtsuki, T. Kokubo, K. Takatsuka, T. Yamamuro, J. Ceram. Soc. Japan (Seramikkusu Ronbunshi) **99**, 1 (1991)
11. C. Ohtsuki, T. Kokubo, T. Yamamuro, J. Non-Cryst. Solids **143**, 84 (1992)
12. W. Bonfield, M.D. Grynpas, A.E. Tully, J. Bowman, J. Abram, Biomaterials **2**, 185 (1981)

13. M. Kikuchi, S. Itoh, S. Ichinose, K. Shinomiya, J. Tanaka, Biomaterials **22**, 1705 (2001)
14. O. Suzuki, Acta Biomater. **6**, 3379 (2010)
15. O. Suzuki, Jpn. Dent. Sci. Rev. **49**, 58 (2013)
16. Y. Hu, J.D. Mackenzie, J. Mater. Sci. **27**, 4415 (1992)
17. K. Tsuru, C. Ohtsuki, A. Osaka, T. Iwamoto, J.D. Mackenzie, J. Mater. Sci. Mater. Med. **8**, 157 (1997)
18. K. Tsuru, S. Hayakawa, C. Ohtsuki, A. Osaka, J. Sol-Gel Sci. Tech. **13**, 237 (1998)
19. T. Yabuta, K. Tsuru, S. Hayakawa, C. Ohtsuki, A. Osaka, J. Sol-Gel Sci. Tech. **19**, 745 (2000)
20. C. Ohtsuki, T. Miyazaki, M. Tanihara, Mater. Sci. Eng. C **22**, 27 (2002)
21. C. Ohtsuki, T. Miyazaki, M. Kyomoto, M. Tanihara, A. Osaka, J. Mater. Sci. Mater. Med. **12**, 895 (2001)
22. P. Li, C. Ohtsuki, T. Kokubo, K. Nakanishi, N. Soga, K. de Groot, J. Biomed. Mater. Res. **28**, 7 (1994)
23. H.-M. Kim, F. Miyaji, T. Kokubo, T. Nakamura, J. Biomed. Mater. Res. **32**, 409 (1996)
24. T. Kokubo, S. Yamaguchi, Acta Biomater. **44**, 16 (2016)
25. C. Ohtsuki, H. Iida, S. Hayakawa, A. Osaka, J. Biomed. Mater. Res. **35**, 39 (1997)
26. S. Kaneko, K. Tsuru, S. Hayakawa, S. Takemoto, C. Ohtsuki, T. Ozaki, H. Ionue, A. Osaka, Biomaterials **22**, 875 (2001)
27. T. Kim, M. Suzuki, C. Ohtsuki, K. Masuda, H. Tamai, E. Watanabe, A. Osaka, H. Moriya, J. Biomed. Mater. Res. Part B: Appl. Biomater. **64B**, 19 (2003)
28. T. Miyazaki, H.-M. Kim, T. Kokubo, C. Ohtsuki, H. Kato, T. Nakamura, Biomaterials **23**, 827 (2002)
29. H. Kato, T. Nakamura, S. Nishiguchi, Y. Matsusue, M. Kobayashi, T. Miyazaki, H.-M. Kim, T. Kokubo, J. Biomed. Mater. Res. **53**, 28 (2000)
30. M. Uchida, H.-M. Kim, T. Kokubo, M. Nawa, T. Asano, K. Tanaka, T. Nakamura, J. Biomed. Mater. Res. **60**, 277 (2002)
31. Y. Sogo, T. Sakurai, K. Onuma, A. Ito, J. Biomed. Mater. Res. **62**, 457 (2002)
32. K. Ishikawa, Y. Miyamoto, T. Yuasa, A. Ito, M. Nagayama, K. Suzuki, Biomaterials **23**, 423 (2002)
33. A. Obata, T. Hotta, T. Wakita, Y. Ota, T. Kasuga, Acta Biomaterialia **6**, 1248 (2010)
34. K. Yamashita, N. Oikawa, T. Umegaki, Chem. Mater. **8**, 2697 (1996)
35. T. Kobayashi, S. Nakamura, K. Yamashita, J. Biomed. Mater. Res. **75**, 477 (2001)
36. H. Yuan, Z. Yang, J.D. de Bruijn, K. de Groot, X. Zhang, Biomaterials **22**, 2617 (2001)
37. M. Kawashita, J. Ceram. Soc. Japan **126**, 1 (2018)

Chapter 7
Mechanical Properties of Ceramics

Junichi Matsushita

Abstract In this chapter, mechanical properties of ceramics are described on the basis of both standards (JIS and ISO) together, for strength, hardness, creep, and abrasion resistance of dense sintered body. As a notable point, the characteristic value of mechanical property of ceramics depends on the size and quantity of microscopic and macroscopic defects (endogenous defect and exogenous one). Furthermore, the anisotropy and crystallinity in microstructure of the sample have an influence on the actual value. Besides, the measurement value is influenced by the measurement condition (different jig and device) even if it is measured with the same instrument and equipment. Nevertheless, the characterization of ceramics requires to evaluate more accurate mechanical property by using an appropriate test method.

Keywords Density · Strength · Hardness · Toughness

7.1 What Is the Mechanical Properties of Ceramics?

In a wide sense, ceramic materials are comprised of inorganic nonmetals excluding metal and alloy. Usually, they are produced in the forms of sintered body, single crystal, amorphous (glass), powder, whisker, fiber, and composite. Raw powders and green compacts in the manufacturing process are also included in general meaning. However, the measurement and evaluation of mechanical properties of these various ceramics are impossible under the same standardized condition. Normally, the mechanical properties of ceramics are defined for dense sintered body. The sintered texture is called microstructure which consists of crystalline grains bonding to each other. The mechanical property seriously depends on the defect structure. The crystalline grain has lattice defects inside (typically, dislocation). The microstructure includes pores if the sintering is not complete and/or foreign inclusion particles if the starting purity is low. Large defects lower the strength. Anisotropic grains affect the

J. Matsushita (✉)
Department of Materials Science, Tokai University, Hiratsuka, Japan
e-mail: matsushita@tsc.u-tokai.ac.jp

J. Hojo (ed.), *Materials Chemistry of Ceramics*,
https://doi.org/10.1007/978-981-13-9935-0_7

crack propagation. Glassy phase weakens the material, particularly at high temperature. The measurement of mechanical property encounters some problems caused by measurement conditions: specimen size, loading rate and pressure, and loading manner. Even if an expensive instrument is used and an appropriate condition is set, the correctness of measured value should be verified with the deep insight on how and why the final value is obtained. For that purpose, the standardized evaluation method of mechanical property of ceramics, particularly fine ceramics, has been prescribed in each country: Japanese Industrial Standards (JIS) in Japan [1]. The JIS is a national standard established on the basis of the industrial standardization method. In late years, the international standardization is the major tendency for the measurement of mechanical property of ceramics, being now led by the specialized committee (technical committee) in the field of fine ceramics for the establishment of the International Organization for Standardization (ISO) standard.

In this chapter, the strength, hardness, creep (transformation dependent at time), fatigue, friction abrasion, etc. of dense sintered body are exposited by using accordingly JIS and ISO standards [2].

7.2 Mechanical Properties of Sintered Body

In the evaluation of mechanical properties of ceramics, particularly dense sintered body, the internal factors inside the sample, which have a large effect on the characteristic values, are constitution phase (crystalline, amorphous), size and shape of grains, grain boundary (crystal orientation, boundary phase), pore (quantity, size), and foreign inclusion (quantity, ingredient). In addition, microscopic defect or distortion has also an influence because ceramics have a rigid crystal structure owing to the strong covalent and ionic bonding of atoms and ions. Ceramics are destructed in a different manner from metals. Metallic materials exhibit a ductile fracture behavior, in which they are destroyed after plastic deformation following to elastic deformation. On the other hand, ceramics are not deformable, and they are catastrophically collapsed in a brittle fracture manner within the elastic deformation, especially at low temperatures. Normally, the measurement of mechanical property is carried out at room temperature and ordinary pressure in ambient air. Meanwhile, various conditions are also used appropriately: gas atmospheres (vacuum, oxidation, reducing, neutral (inert)), liquid atmospheres (water, solution, melt), low and high temperatures, and low and high pressure [2].

7.2.1 *Density and Porosity*

There are various definitions of density such as bulk density, theoretical density, true density, apparent density, and relative density even if it is called "density" in one word. The packing density of fine ceramics powder is defined as particle density and bulk density, and the measurement methods are prescribed by JIS R 1620 and JIS R 1628, respectively, in Japan. The measurement methods of bulk density and open porosity of fine ceramics sintered body are prescribed in JIS R 1634 [3]. In ISO, the measurement methods of various densities are standardized as "specified methods for the determination of apparent solid density, bulk density, apparent density and geometric bulk density of fine ceramics" (ISO 18754).

Generally, mechanical property of dense sintered body is sensitively affected by distribution, size, and quantity of pores. Thus, the influence of density and porosity cannot be ignored when mechanical property is evaluated. Even if sintered bodies are almost the same in composition, bulk density, and porosity, the difference in grain size, inclusion, and grain boundary between sintered textures also influences the mechanical property. Therefore, the most suitable fabrication process is important to control various factors on mechanical properties. Anyway, density and porosity of test specimen should be checked as principle factors for evaluation of mechanical property. The bulk density of sintered body is measured by a liquid immersion method using the principle of floatation, which is popularly called the Archimedes' method. When the specimen is weighed in a liquid, the measured weight is small compared to the real weight. From the difference, the volume of specimen is calculated. When using this method, open pores must be perfectly filled by the immersion liquid. It is desirable to use a big size specimen to lower the measurement error and to use multiple specimens if they are obtained from the same lot. Bulk density (ρ, g/cm^3) and open porosity (P, %) are calculated using Eqs. 7.1 and 7.2.

$$\rho = (W_1/(W_3 - W_2)) \times \rho_1 \tag{7.1}$$

$$P = ((W_3 - W_1)/(W_3 - W_2)) \times 100 \tag{7.2}$$

where W_1 is dry mass (g), W_2 is mass in liquid (g), W_3 is mass of liquid saturated specimen (g), and ρ_1 is density of liquid (g/cm^3). Normally, water is used as immersion liquid at room temperature.

Bulk density includes open porosity and closed porosity. Apparent density includes closed porosity without open porosity. True density is the real density without open and closed porosities. Theoretical density is calculated from crystallographic data obtained by X-ray diffraction analysis. Relative density is obtained by dividing bulk density by theoretical density or true density (Eq. 7.3).

$$RD = (BD/TD) \times 100 \tag{7.3}$$

where RD is relative density (%), BD is bulk density (g/cm^3), and TD is theoretical density (g/cm^3).

The theoretical density can be easily calculated for monolithic sintered body and simple mixed type of composite but is difficult to calculate when the composition and quantity of constituent phases are not clear by formation of solid solution and compounds.

In such a case, the true density is measured after fine grinding of the sintered body. However, when the quantity of formed phase is small, relative density is allowed to calculate on the assumption of simple mixture for relative comparison [2, 4].

7.2.2 Hardness

A traditional evaluation of hardness is the Mohs hardness. This is a relative comparison of hardness by scratch test and has been used for natural ores. Conventional hardness test is an indentation type: the Vickers hardness (abbr. HV (VHN)), the Knoop hardness (abbr. HK), the Rockwell hardness (abbr. HR, HRA, HRC scales, etc.), the Brinell hardness (abbr. HB), micro hardness (micro Vickers hardness, micro Knoop hardness), etc. The Shore hardness (abbr. HS) is a rebounding type.

Generally, the hardness of dense sintered body is evaluated using the Vickers indentation. The Knoop hardness is often used for especially brittle materials such as inorganic glass. The Shore hardness is often used for carbon-based materials. A hardness test of fine ceramics at room temperature and high temperature in JIS is the Vickers indentation. In the ISO standard, the Vickers hardness and the Knoop hardness are prescribed for monolithic ceramics at room temperature.

The Vickers hardness test uses a quadrangular pyramidal indenter made of diamond single crystal with the opposite face angle of 136°. The indenter is pressed to the specimen surface under a constant-speed loading. After then, the hardness is calculated from the diagonal length of indentation shown in Fig. 7.1 [4], by using Eq. 7.4. Finally, the measured value is converted to the International System of Units, SI (GPa).

$$Hv = 0.12\ F/S = 0.1022\ F\ \sin\theta/2/d^2 = 0.1891\ F/d^2 \tag{7.4}$$

where Hv is the Vickers hardness, F is load (N), S is surface area of indentation (mm^2), d is mean diagonal length of indentation (mm), and θ is opposite face angle of pyramidal indenter.

The Vickers hardness test is usually conducted with a load of 1 kgf (9.807 N) to 20 kgf (196.14 N). The standard load in JIS is 1 kgf (9.807 N) and 10 kgf (98.07 N). Since the indentation shape is homothetic for loading, the measured hardness does not change theoretically even if the load is changed. However, the size of indentation tends to become small for high hardness materials at the same load, resulting in a large error in measurement of diagonal length. Generally, a high load is imposed within the load range not to break near the edge of indentation.

The Vickers hardness of various materials is shown in Table 7.1 [4]. As a representative metal, the Vickers hardness of steel (Fe) and pure aluminum (Al) is about 2 GPa and 0.3 GPa, respectively, at room temperature. As a hard ceramics, the Vickers hardness of B_4C and TiB_2 is extremely high as about 35 GPa and 32 GPa, respectively, at room temperature. The load level of micro Vickers hardness (micro

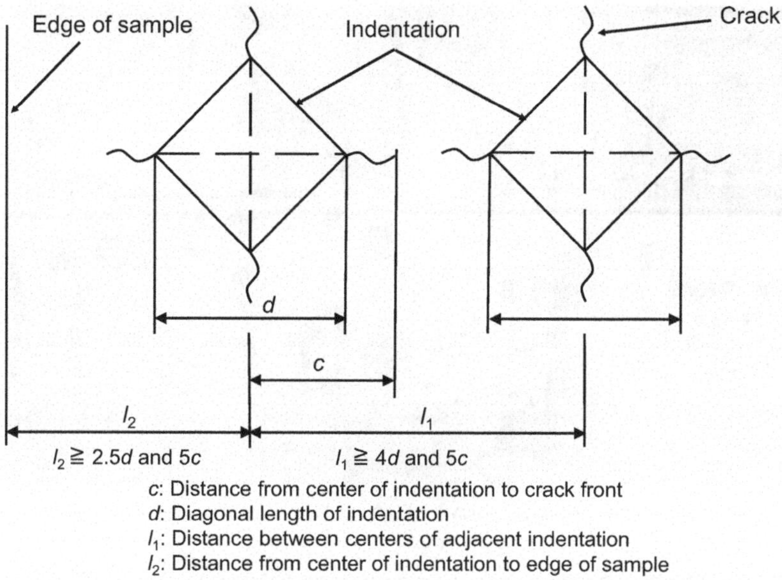

Fig. 7.1 Schematic diagram of indentation and crack on the surface of ceramics by the Vickers indenter. (Modified with permission from Ref. [4], Copyright 2005 The Chemical Society of Japan)

Table 7.1 The Vickers hardness of various materials

Group	Materials	Hardness, H_v (GPa)
Ceramics	α-Al_2O_3	18 ~ 20
	c-ZrO_2	12 ~ 13
	B_4C	30 ~ 35
	α-SiC	22 ~ 24
	h-BN	28 ~ 45
	β-Si_3N_4	14 ~ 18
	TiB_2	32 ~ 34
	SiB_6	18 ~ 22
	C (diamond)	80 ~ 100
Metals	Au	0.22
	Ag	0.24
	Cu	0.5 ~ 1
	Fe (steel)	1 ~ 3
	Al	0.2 ~ 0.4
	Ti	1 ~ 3

Modified with permission from Ref. [4], Copyright 2005 The Chemical Society of Japan

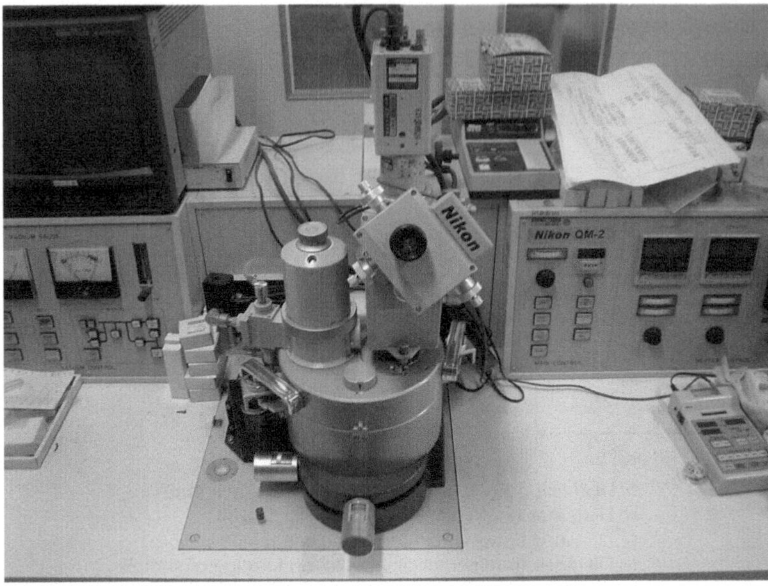

Fig. 7.2 Example of testing equipment for the micro Vickers hardness

hardness) measurement in JIS is from 1 gf (9.807 mN) to 1 kgf (9.807 N). The Vickers hardness test of porous material should avoid the low-load condition, because pores disturb the indentation in the small area.

Figure 7.2 demonstrates the equipment for the micro Vickers hardness test which is available in various atmospheres from room temperature to high temperature region over 1000 °C. Figure 7.3 [4] shows the temperature dependence of the micro Vickers hardness of various ceramics measured using this equipment in vacuum. The micro Vickers hardness of Ni alloy and Cr steel is usually about 2.5 GPa and 4 GPa, respectively, at room temperature. The micro Vickers hardness of α-Al_2O_3 and α-SiC, which are widely used as engineering ceramics, is high as about 18 GPa and 24 GPa, respectively, at room temperature and is still high at high temperatures.

The Knoop hardness test is similar to the Vickers hardness test but uses a rhomboid pyramidal indenter made of diamond single crystal with opposite face angles of 172°30' and 130° in long and short axes, respectively. The Knoop indenter is pressed on to the specimen surface at a constant-speed loading, and the Knoop hardness is calculated from the diagonal length at the long axis.

The Rockwell hardness test uses a steel ball or a diamond cone. They are pressed on the specimen surface at a constant load, and the hardness is calculated from the indentation depth. The Brinell hardness test uses an alloy ball or a diamond ball, and the hardness is calculated from the surface area of indentation. The Rockwell hardness and the Brinell hardness are used for the evaluation of hardness of metal or plastic.

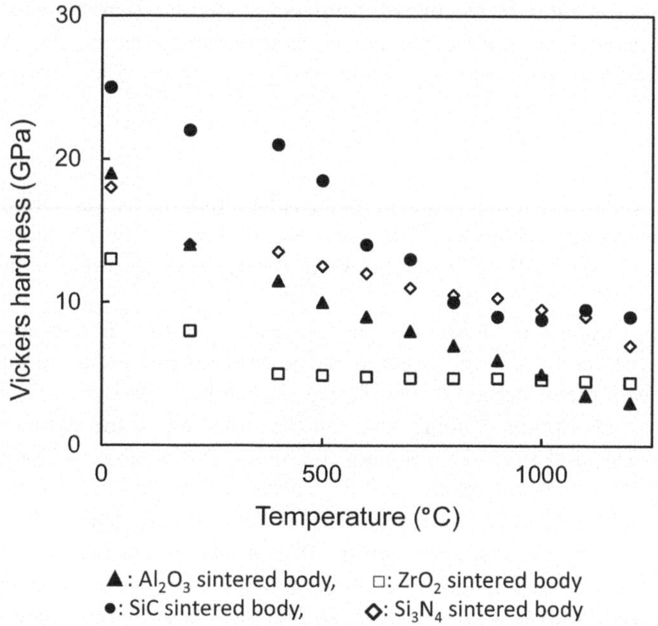

Fig. 7.3 The micro Vickers hardness of various ceramics at high temperature. (Modified with permission from Ref. [4], Copyright 2005 The Chemical Society of Japan)

7.2.3 Elastic Modulus

Elastic modulus is the stress/strain ratio within the elastic deformation. The Young's modulus, modulus of rigidity (shear modulus), and the Poisson's ratio are included as mechanical property of ceramics. The elastic modulus is the measure of resistance against the external force working to separate ions and atoms chemically bonded in ceramics. There are static method and dynamic method for the evaluation.

The Young's modulus is the reciprocal value of proportional constant in relationship between elongation per unit length and tension per unit cross section when a material is pulled. On the other hand, the shear modulus is a proportional constant between shear stress and shear strain. Tentatively now, a low external stress is burdened to an isotropic elastic body. The proportional relationship between stress and strain is expressed in Eqs. 7.5 and 7.6 (The Hooke's law).

$$\sigma = E\varepsilon \tag{7.5}$$

where σ is perpendicular stress, E is the Young's modulus, and ε is perpendicular strain.

$$\tau = G\gamma \tag{7.6}$$

where τ is shear stress, G is shear modulus, and γ is shear strain.

The Poisson's ratio is calculated from perpendicular (lengthwise) strain and horizontal (crosswise) strain generated by a stress in perpendicular (lengthwise) direction (Eq. 7.7).

$$\nu = -\varepsilon_z / \varepsilon_y \tag{7.7}$$

where ν is the Poisson's ratio, ε_z is perpendicular strain, and ε_y is horizontal strain.

The measurement methods of the static and dynamic Young's moduli at room temperature are standardized in JIS, as well as the dynamic Young's modulus at high temperature.

The static Young's modulus (isothermal elastic modulus) is measured from the relationship between stress and strain in elastic deformation under the static load. In JIS, the measurement method of the Young's modulus is prescribed using 3-point bending test and 4-point bending test. On the other hand, the dynamic Young's modulus is measured from the resonant frequency under forced vibration of plate specimen or from the propagation speed of ultrasonic wave pulse inside the specimen excited at high frequency. The former is named the resonance method (bending resonance method). The test specimen in JIS is a rectangular parallelepiped with a dimension over 40 mm in length, over 5 mm in width, and over 1 mm in thickness, in which the length/thickness ratio is over 20. The latter is called the supersonic wave pulse method. The test specimen in JIS is a rectangular column of 10 mm square or a cylindrical column of 10 mm diameter and over 20 mm length.

Generally, the important concept of product design in engineering ceramics is light weight and thin wall thickness. In this case, high Young's modulus ceramics is preferable to attain the small deformation by external pressure. Meanwhile, since low Young's modulus ceramics is generally small in thermal stress evolution, they are used under high thermal stress condition. The Young's moduli of various materials are shown in Table 7.2 [4]. As representative polymer, the Young's moduli of polyethylene and nylon 6 are about 1 GPa and 2 GPa, respectively, at room temperature, although they depend on measurement condition. As representative metal, the Young's moduli of steel (Fe) and pure aluminum (Al) are about 200 GPa and 70 GPa, respectively. As representative engineering ceramics, the Young's moduli of dense alumina sintered body (α-Al_2O_3) and silicon carbide sintered body (α-SiC) are high as about 400 GPa and about 480 GPa, respectively, at room temperature.

7.2.4 Strength

There are various tests of stress evaluation, that is, strength evaluation on the fracture of ceramics as follows: bending strength (flexural strength), compressive strength, tensile strength, and shock strength. The strength value of sintered body depends on the test method and loading condition. Ceramics cause the catastrophic fracture

Table 7.2 The Young's modulus of various materials

Group	Materials	Young's modulus (GPa)
Ceramics	α-Al_2O_3	380 ~ 400
	c-ZrO_2	210
	B_4C	450
	α-SiC	400 ~ 490
	h-BN	700 ~ 900
	β-Si_3N_4	220 ~ 310
	TiB_2	450
	ZrB_2	300 ~ 345
	C (diamond[a])	1210
Metals	Au	78
	Ag	83
	Cu	129
	Fe (steel)	211
	Al	70
	Ti	116
Polymers	Nylon 6	2.8
	Polyethylene	0.4 ~ 1.3
	Polypropylene	1.4

[a]Orientation (111)

Modified with permission from Ref. [4], Copyright 2005 The Chemical Society of Japan

without yielding at room temperature by excessive load-like stress concentration (brittle fracture). Ceramics have a rigid structure constituted by strong covalent bond and ionic bond of cation and anion with different ionic sizes. Therefore, the movement and growth of dislocation are difficult, generating the stress concentration near the defects (voids, cracks, and grain corners). Meanwhile, metals are ruptured after stress is relaxed by yielding on excessive load-like stress concentration (ductile fracture).

Typical stress-strain diagrams (stress-strain curves) of ceramics and metals at room temperature are illustrated schematically in Fig. 7.4 [4]. As for ceramics, the linear relationship is indicated between stress (σ) and strain (ε) up to fracture point (σ_c). On the other hand, as for metals, the linear relationship is indicated up to yielding point (σ_y), and then the stress reaches the maximum point (σ_{max}) via non-linear region. After then, the necking by constriction starts and the stress decreases, finally leading to rupture at the ductile fracture point (σ_m). Here, an ideal ceramic crystal is assumed for brittle fracture. The stress required for fracture of crystal is equal to that for generation of new surface. Simply stated, it is the stress required to break the bond between atoms. Supposing a sinusoidal relationship between stress and strain, Eq. 7.8 was introduced by Orowan in the USA. σ_{th} in Eq. 7.8 is the critical fracture stress which is called ideal strength or theoretical strength [5].

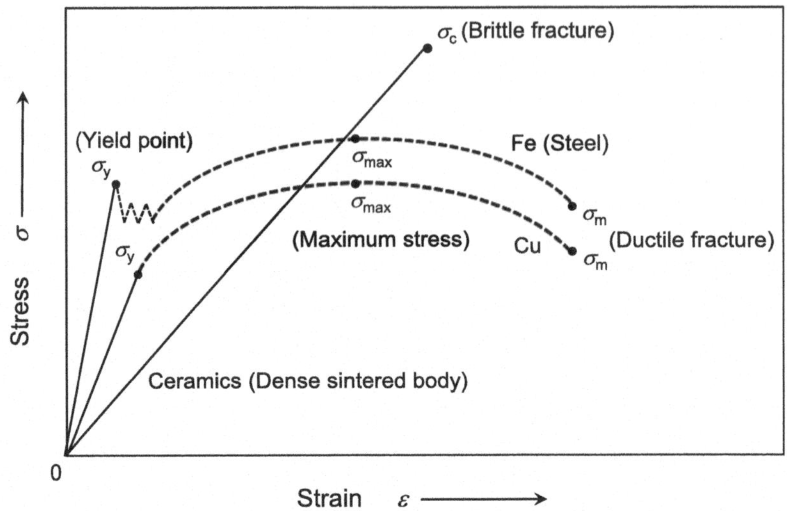

Fig. 7.4 Stress-strain curves of ceramics and metals at room temperature. (Modified with permission from Ref. [4], Copyright 2005 The Chemical Society of Japan)

Table 7.3 Ideal strength and practical one of typical ceramics

Materials	Ideal strength (GPa)	Strength of sintered body (GPa)
α-Al_2O_3	40	0.4
c-ZrO_2	14	1.2
α-SiC	22	0.8
β-Si_3N_4	10	1.0

Modified with permission from Ref. [4], Copyright 2005 The Chemical Society of Japan

$$\sigma_{th} = (\gamma_0 E/a_0)^{1/2} \tag{7.8}$$

where σ_{th} is ideal strength; γ_0 is energy per unit area to produce fracture surface (new surface), that is, surface energy of fracture surface; E is the Young's modulus; and a_0 is atomic distance of fracture surface.

Considering based on Eq. 7.8, the ideal strength of ceramics is high when surface energy of fracture surface and the Young's modulus are large and atomic distance is short. The ideal strength of typical ceramics is shown in Table 7.3 [4]. The ideal strength level is 100–1000 times higher than an actual strength level of practical ceramic sintered bodies. The significantly large gap between theoretical and actual values is caused by stress concentration into defects of practical ceramics, that is, defects are weak points as fracture origin.

The types of loading to evaluate the strength of ceramics are bending mode, tensile mode, compressive mode, and torsional mode. An example of strength test equipment is shown in Fig. 7.5. The strength test of ceramics is prescribed in JIS and ISO as follows: bending test (room temperature, high temperature), tensile test (room temperature, high temperature), and compressive test (room temperature). As

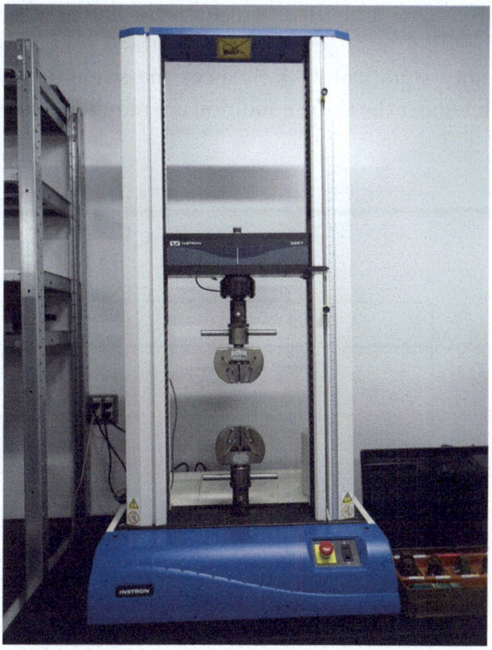

Fig. 7.5 Example of strength testing equipment

strength test of ceramics, the bending test is common. The main reason is a small size and an easy preparation of specimen compared to tensile test etc. Impact strength is not classified into strength evaluation but into fracture toughness evaluation. There are the Charpy impact test and the Izot impact test, but they are not common as strength test of ceramics and not standardized by JIS.

The bending strength is measured by 3-point method (Fig. 7.6 (a) [4]) and 4-point method (Fig. 7.6 (b) [4]) and calculated according to Eqs. 7.9 and 7.10, respectively.

$$\sigma_{b3} = 3PL/2wt^2 \tag{7.9}$$

$$\sigma_{b4} = 3P(L - l)/2wt^2 \tag{7.10}$$

where σ_{b3} is 3-point bending strength (N/mm^2), σ_{b4} is 4-point bending strength (N/mm^2), P is the maximum load at fracture point (N), L is under support span (mm), l is upper loading span (mm), w is width of specimen, and t is thickness of specimen (mm).

The dimension of specimen for 3-point and 4-point tests is over 36 mm in length (normally 40–50 mm), 4 mm in width, and 3 mm in thickness (Fig. 7.6c [4]). The under support span in 3-point bending test is 30 mm. The upper loading span and the under support span in 4-point bending test are 10 mm and 30 mm, respectively. The perfect joint-type support (freely rotatable bearing jig) must be used as the under support and the upper loading support in ISO (Fig. 7.7 [4]). The preparation of bending test specimen by machining should pay attention to chamfering (rounding of ridge) and surface roughness and damage by machining. Particularly, as for bending test specimen, the surface roughness on tensile stress side (under span side) significantly influences the strength value. Multiple tests by using the same lot under the same condition are required to confirm the fluctuation of data in order to

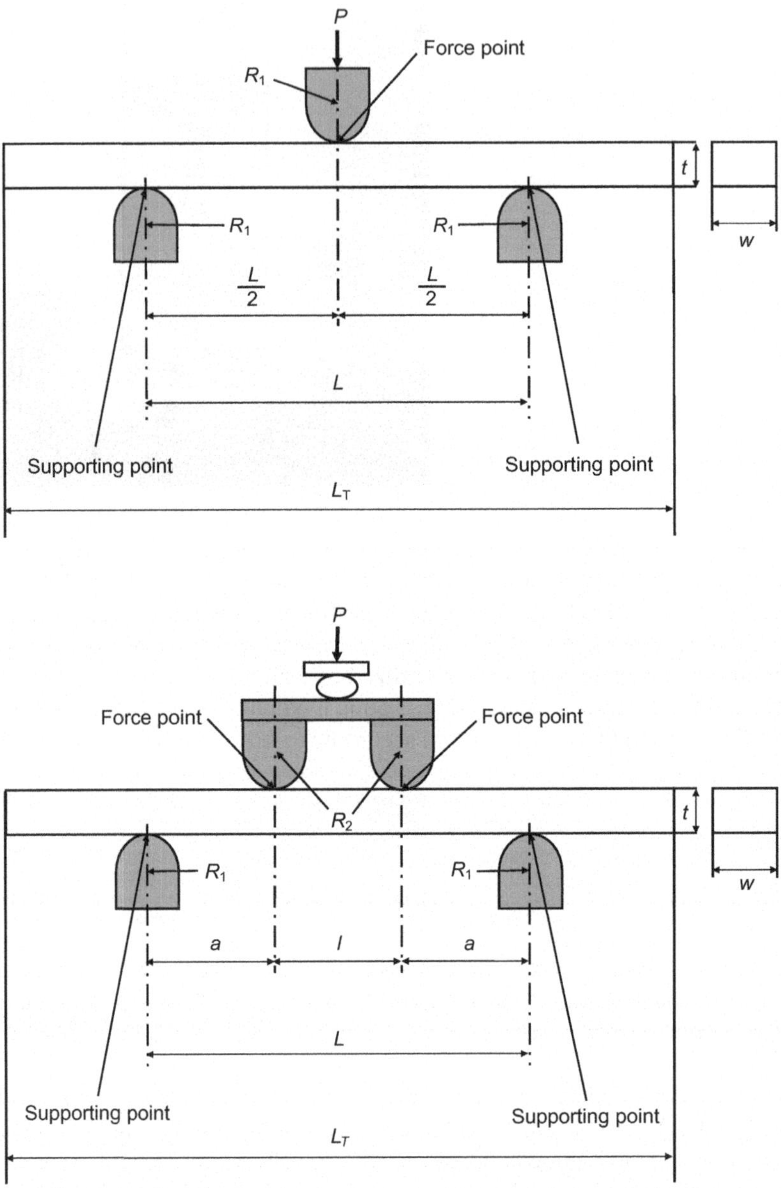

Fig. 7.6 Schematic diagram of bending test by 3-point method and 4-point method. (Modified with permission from Ref. [4], Copyright 2005 The Chemical Society of Japan)

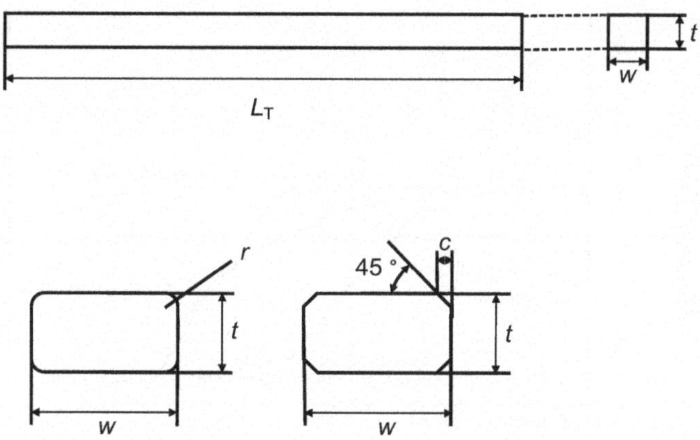

Fig. 7.6 (continued)

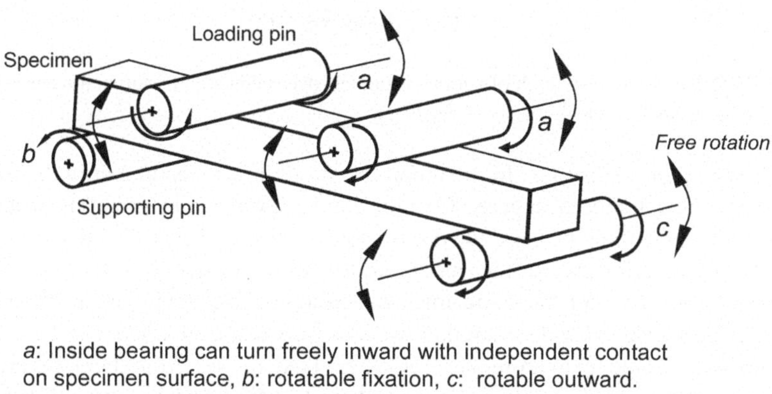

Fig. 7.7 Schematic diagram of perfect joint-type support. (Modified with permission from Ref. [4], Copyright 2005 The Chemical Society of Japan)

assess whether the measured value is the characteristic one of material or underestimated by failure machining.

Tensile strength test is used to evaluate particularly the plastic deformation or the super plasticity at high temperature. The reason is because the constant-rate tensile loading up and down is easy in the soaking area of gauge part (constriction part) of the specimen. In bending test, the deflection of specimen may deviate from soaking area in furnace. To compare the strength of ceramics, more exactly, the specimen machined with the same size under the same condition should be used. Ceramics are sensitive to impact load and stress concentration and strongly influenced by defects and flaw. The existing flaw, segregating phase and pore as fracture origin, should be controlled at the same level to compare the strength. Although the measured strength should be the same on calculation regardless of the specimen size, the strength value

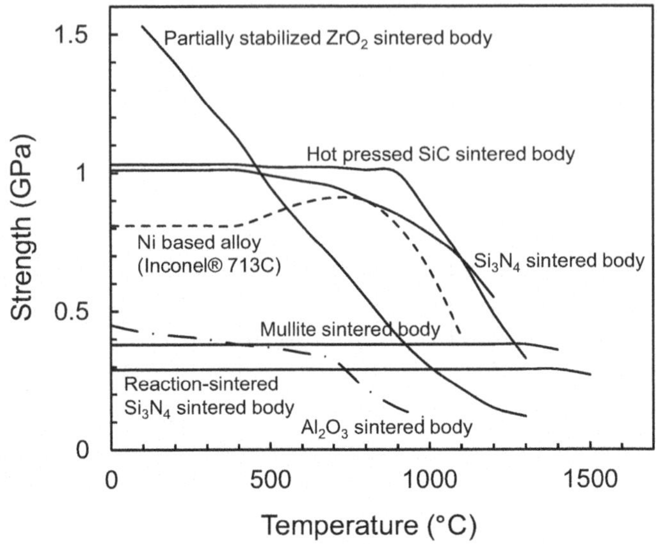

Fig. 7.8 High temperature strength of various ceramics. (Modified with permission from Ref. [4], Copyright 2005 The Chemical Society of Japan)

of small specimen tends to be high. When the strength is measured at high temperature over a few hundred degrees Celsius, the material of jig (bending test tool) should receive attention, because the plastic deformation and chemical reaction of jig should be avoided for precise measurement. Since the mismatch of thermal expansion coefficient between jig and specimen causes a problem, the jig made of the same material is desirable. Metallic materials such as heat-resistant alloys can be used in the temperature range up to around 800 °C (e.g., TiAl alloys). At higher temperature, dense sintered silicon carbide (α-SiC) or mullite ($3Al_2O_3{\cdot}2SiO_2$) is generally used as material of jig.

The high-temperature strength of various ceramics is shown in Fig. 7.8 [4]. These bending strength values are reference ones and can be improved in room temperature strength and resistance against reduction of strength at high temperature by control of production process and resulting microstructure (porosity, impurity, grain size, and grain boundary). However, the strength tends to decrease with increasing test temperature over 1000 °C because of deterioration of crystal components themselves or softening of grain boundary in sintered ceramics. From this point of view, non-oxide ceramics such as silicon carbide (SiC) and silicon nitride (Si_3N_4) with high strength at high temperature are widely used as high temperature structural material [6–10].

7.2.5 Fracture Toughness

"Toughness" means the tenacity of material, and the tenacity against destruction is called "fracture toughness." The fracture toughness is a typical mechanical property of ceramics and represents the measure of "brittleness" of ceramics. Stress intensity factor (K) is used to evaluate the fracture toughness and represents the intensity of stress field near the tip of crack, where the specific stress field becomes infinite at the crack front. This factor K_{IC} has been developed by Griffith, in the UK in 1920 [11], and Irwin, in the USA in 1957 [12]. Griffith assumed the loading of uniaxial tensile stress to a ceramics thin plate with a defect of $2C$ long as shown in a model (Fig. 7.9 [4]). When the stress concentrates on this defect and exceeds a certain stress level, a crack progresses, leading to fracture. According to this assumption, Eq. 7.11 was derived, and Eq. 7.12 was introduced. The factor K is calculated from crack length, external stress, and specimen shape. The critical value (K_C) of the stress intensity factor (K) is named a fracture toughness value. Ceramics have a resistance against tensile stress and shear stress, and the displacement type (mode) at the crack tip surface for crack propagation (crack growth) is classified into the following three modes (Fig. 7.10 [4]): tensile opening mode (mode I), sliding in-plane shear mode (mode II), and tearing out-plane (anti-plane) shear mode (mode III). Each mode (I, II, III) is independent from each other, but, of course, the destruction in the mixed mode is also possible. Generally, the critical stress intensity factor of ceramics is evaluated in mode I, where the resistance against crack propagation works on the tensile stress in a crack surface. The K_{IC} means the critical stress intensity factor in mode I [13].

$$\sigma_f = (2E\gamma/\pi C)^{1/2} \tag{7.11}$$

where σ_f is fracture stress, E is the Young's modulus, γ is surface energy, π is circle ratio, and C is crack radius.

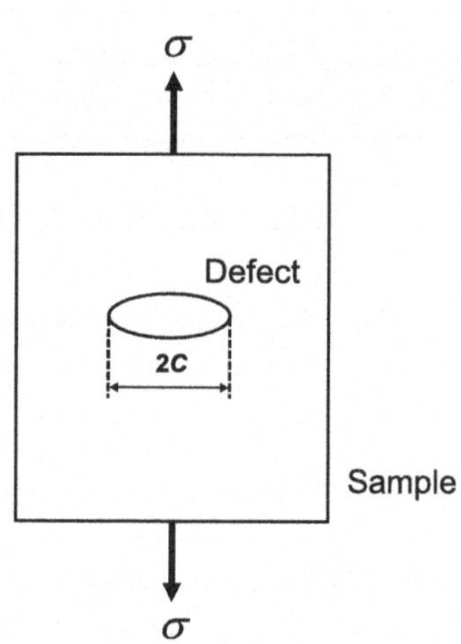

Fig. 7.9 Schematic diagram of defect model in ceramics. (Modified with permission from Ref. [4], Copyright 2005 The Chemical Society of Japan)

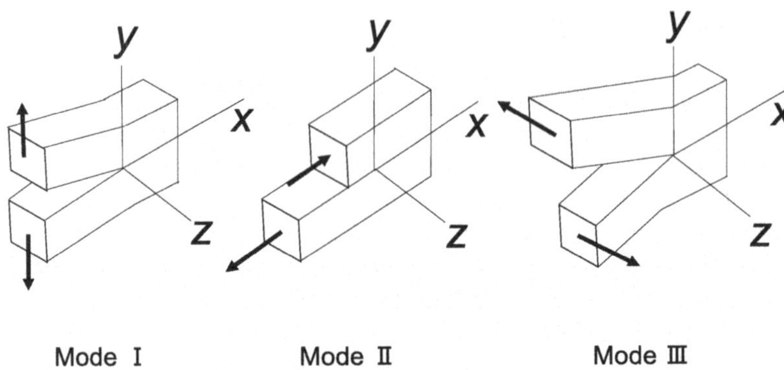

Fig. 7.10 Schematic diagram of surface displacement type of crack tip. (Modified with permission from Ref. [4], Copyright 2005 The Chemical Society of Japan)

$$K_{IC} = (EG_c)^{1/2} \tag{7.12}$$

where K_{IC} is fracture toughness, E is the Young's modulus, and G_C is the critical value of crack extension force.

The test methods of fracture toughness (K_{IC}) of ceramics are proposed as follows: the notched-beamed method, the chevron-notched-beamed method (popular name: CNB method), the controlled-surface-flow method (popular name: CSF method), the single-edge-precracked-beam method (popular name: SEPB method), the single-edge V-notch beam method (popular name: SEVNB method), the surface crack in flexure method (popular name: SCF method), and the indentation fracture method (popular name: IF method). In JIS, two kinds of pre-crack introduced method and indentation method are prescribed as fracture toughness (K_{IC}) test method of fine ceramics at room temperature. As for fracture toughness test at high temperature, pre-crack introduced method is prescribed. In ISO, SEPB method, SEVNB method, CSF method, CNB method, and SCF method are standardized as fracture toughness (K_{IC}) test for monolithic ceramics. Fracture toughness of typical ceramics is shown in Table 7.4 [4] in comparison with other materials.

7.2.6 Thermal Shock Resistance

When ceramics are rapidly heated or cooled, temperature distribution is generated between the surface and the inside of specimen, as a result, inducing thermal stress. When the specimen is quenched, the surface thermally shrinks more than the inside, generating the tensile stress on the surface. This is the cause of thermal shock rupture. Thermal shock resistance test is performed to estimate the critical thermal stress, that is, how much thermal stress and how much temperature difference in rapid heating or quenching cause the fracture of specimen. In JIS, the thermal shock resistance test by water quenching method using bending strength test is prescribed

Table 7.4 Fracture toughness of typical ceramics

Materials[a]	K_{IC} (MPam$^{1/2}$)
α-Al_2O_3	2.5 ~ 4.5
c-ZrO_2	8 ~ 12
B_4C	4 ~ 6
α-SiC	2.5 ~ 4.5
β-Si_3N_4	4 ~ 9
AlN	3 ~ 4
TiB_2	3 ~ 5
SiB_6	3 ~ 5
C (diamond)[b]	2 ~ 2.5

[a]Dense-sintered body
[b]Single crystal, orientation (111)
Modified with permission from Ref. [4], Copyright 2005 The Chemical Society of Japan

for fine ceramics. The standard test piece is column-shaped with a length of 50–100 mm, and the top of column must be conically processed. Ordinarily, the thermal shock resistance is evaluated by water quenching method using specimen for bending strength test.

As an example, dense sintered Al_2O_3 specimen (size: 3 mm × 4 mm × 40 mm) held at a certain temperature was dropped into water at 0 °C. After then, 4-point bending strength of quenched specimen was measured as shown in Fig. 7.11 [4]. The quenched specimen maintained the strength of initial specimen up to the temperature difference of 200 °C. Then, the strength decreased drastically at the temperature difference of 250 °C because cracks form in the specimen by thermal shock stress. From this result, the thermal shock resistance (ΔT) of this Al_2O_3 sintered body can be estimated to be approximately 200 °C. This test method is conventionally used for the evaluation of thermal shock resistance of dense sintered body but varies with Biot's modulus depending on the specimen size. The thermal shock resistance is affected by various parameters: fracture strength, the Poisson's ratio, the Young's modulus, thermal expansion coefficient, and thermal conductivity. Biot's modulus represents the vigorousness of thermal shock and depends on specimen size, thermal conductivity, and heat transfer coefficient. Therefore, this conventional method should be used as relative comparative test among materials. White or transparent specimen can be conveniently evaluated by penetrant flaw inspection, in which the quenched specimen is immersed in red ink etc., and the penetrated cracks are observed for estimation of thermal shock resistance.

7.2.7 *Creep and Fatigue*

Creep (time-dependent deformation) is the deformation behavior of material by loading of a constant stress for a long time under a certain environment. There are the three kinds of creep tests: bending creep, tensile creep, and compressive creep.

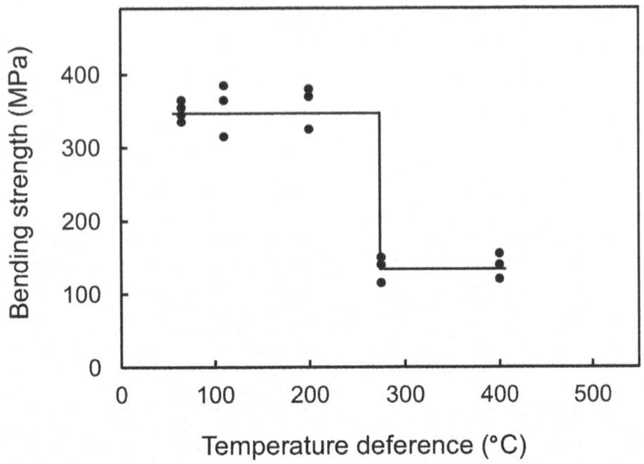

Fig. 7.11 4-point bending strength of quenched Al_2O_3 specimen. (Modified with permission from Ref. [4], Copyright 2005 The Chemical Society of Japan)

Creep test methods of fine ceramics by 4-point bending test and tensile test are prescribed in JIS.

Fatigue is the weakening of material under cyclic stress loading. It is classified into mechanical fatigue under repeated mechanical stress loading and thermal fatigue under repeated thermal stress loading by heating and cooling. There is also thermomechanical fatigue in complex with both mechanical and thermal stresses. The bending fatigue test of fine ceramics at room temperature is prescribed in JIS.

7.2.8 Probability Distribution and the Weibull Distribution

Since brittle material such as ceramics induces the catastrophic fracture instantaneously, it is difficult to exactly estimate the fracture life and to predict the quality guarantee. However, it is important to digitize the probability distribution for the fracture of ceramics as a guide. Above all, the Weibull distribution is popular as statistical model for life data of materials.

In JIS, the Weibull statistical analysis method of the strength data of ceramics is prescribed in a single mode but not in a competing mode. In ISO, "the Weibull statistics for strength data" is prescribed as a probability distribution test method. The Weibull distribution was proposed to predict the fracture life on non-normal probability distribution by Weibull in 1939, Sweden. The Weibull distribution matches well to the distribution of fracture life and is widely used in the world. For example, using the Weibull probability paper, the fracture probability (survival probability, %) and the stress (bending strength) are plotted on the vertical axis and on the horizontal axis, respectively, which is called the Weibull plot, providing the slope (m) of the straight line. The reliability (fracture life) of materials can be judged to be high when the m is large. The m is called the Weibull coefficient [14, 15].

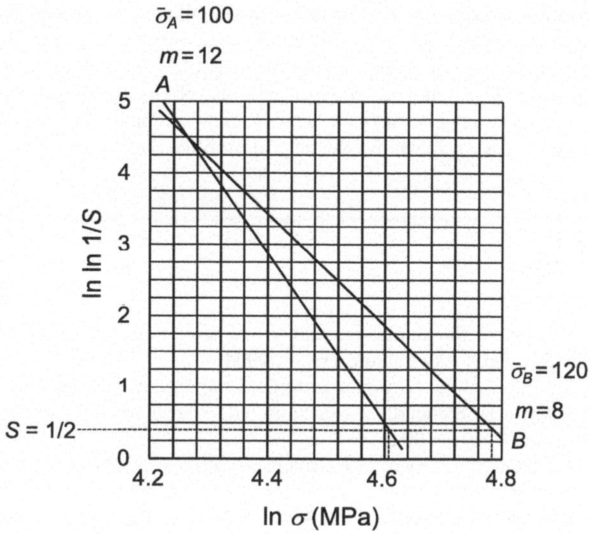

Fig. 7.12 The Weibull plots of two kinds of ceramics. (Modified with permission from Ref. [4], Copyright 2005 The Chemical Society of Japan)

The Weibull plots of two kinds of ceramics (A and B) are shown in Fig. 7.12 [4, 14]. The average strength is given at $S = 1/2$. Ceramics A has the average 4-point bending strength of 100 MPa and the Weibull coefficient of 12, while ceramics B has the average strength of 120 MPa and the Weibull coefficient of 8. From this result, ceramics B is superior to ceramics A in the average bending strength. However, if they are used under the load condition below 69 MPa, ceramics A has the fracture probability less than 1% and is judged to be superior to ceramics B in reliability.

7.2.9 *Friction and Wear Resistance*

The friction and wear become the problems in dynamic contact between materials. The evaluation of friction coefficient and wear resistance is dependent on the combination between the same materials or between the different materials and affected by the external factors such as temperature, atmosphere, and adsorbed water on the specimen surface. Friction and wear properties are usually tested by pin-on-disk method, ball-on-disk method, Okoshi-type method, Tsuya-type method, and Amsler-type method in Japan. In JIS, the abrasion test by ball-on-disk method is prescribed. Also, the ball-on-disk method for friction and wear characteristics is standardized in ISO.

Acknowledgments No chapter of this book may be reproduced in any form without the permission to modify the tables and figures of Ceramics Material Chemistry, pp. 135–158 by Junichi Matsushita edition of the Chemical Society of Japan, Tokyo, Japan [4]. Grateful acknowledgment is made to Ms. Yuko Shimazawa and Ms. Norie Tanaka of the Chemical Society of Japan for permission support. Also, the authors acknowledge the kindness of Ms. Asami Komada of Springer Nature, Tokyo, Japan, for the helpful proofreading suggestion.

References

1. Japanese Standards Association. ed., *International Standardization, JIS Handbook* (Japanese Standards Association, Tokyo, 2004)
2. J. Matsushita, *Introduction to Ceramic Materials* (Baifukan, Tokyo, 2002)
3. Japanese Standards Association. ed., *Ceramics, JIS Handbook* (Japanese Standards Association, Tokyo, 2004)
4. Chem. Soc. Japan, in *Ceramics Material Chemistry*, ed. by J. Hojo et al., (Maruzen, Tokyo, 2005)
5. E. Orowan, Reports on Progress in Physics **12**, 185–232 (1949)
6. Ceram. Soc. Japan ed., *Evaluation Method of Ceramics* (The Ceramic Society of Japan, Tokyo, 1993)
7. T. Nishida, E. Yasuda (eds.), *The Evaluation of Mechanical Property of Ceramics* (The Nikkan Kogyo Shimbun, Tokyo, 1995)
8. H. Awaji, *Strength of Ceramic Materials* (Corona Publishing, Tokyo, 2001)
9. O. Kamigaito, N. Kamiya, *Physics of Ceramics* (Uchida Rokakuho, Tokyo, 1998)
10. Y. Ikuhara (ed.), *Physics of Ceramic Materials* (The Nikkan Kogyo Shimbun, Tokyo, 1999)
11. A.A. Griffith, Philos Trans R Soc A **221**, 163–198 (1920)
12. G.R. Irwin, J. Appl. Mechanics **24**, 361–364 (1957)
13. A.G. Evans, ASTM Spec Tech Publ **678**, 112–135 (1979)
14. H. Makabe, *How to Use Weibull Probability Paper* (Japanese Standards Association, Tokyo, 1995)
15. Y. Matsuo, Translated, ed., *Life and Fracture of Ceramics* (Uchida Rokakuho, Tokyo, 1989)

Further Readings

1. J. Watanabe et al., *Fundamental Materials Engineering* (Kyoritsu Shuppan, Tokyo, 1992)
2. Ceram. Soc. Japan ed., Ceramics Japan **39**, 370–397 (2004)
3. The 156 Committee, JSPS, Japan ed., *Materials System* (Kyoritsu Shuppan, Tokyo, 1997)
4. Ceram. Soc. Japan ed., *Mechanical Property of Ceramics* (Kyoritsu Shuppan, Tokyo, 1979)
5. N. Igata (ed.), *Fundamentals Engineering of Metal Materials* (The Nikkan Kogyo Shimbun, Tokyo, 1995), pp. 113–114
6. H. Okuda et al. (eds.), *Engineering Ceramic Materials* (Ohmsha, Tokyo, 1987)
7. O. Yanagisawa, F. Yoshida, *Fundamentals of Materials Science (New Revision Aufl.)* (Kyoritsu Shuppan, Tokyo, 1998)
8. Y. Moriyoshi et al., *Fundamentals Science of Ceramics* (Uchida Rokakuho, Tokyo, 1993)
9. C.H. Pai, K. Koumoto, *Ceramics* (Dainippon Tosho, Tokyo, 1996)
10. S. Mizuta, K. Koumoto, *Ceramics Materials Science* (University Tokyo Press, Tokyo, 1996)
11. H. Abe et al., *Engineering Ceramics* (Gihodo Shuppan, Tokyo, 1984)

Chapter 8
Electromagnetic Properties of Ceramics

Akira Kishimoto

Abstract Electric properties of ceramics range from insulator to superconductor, which are basically based on their crystal structures. Defect structures such as vacancy and substituted atoms affect some kind of electric properties. Intentional doping is frequently conducted to improve the electric properties. Different from the single crystal, ceramics contain grain boundary giving some unique functions. The conductivity or power generation of some kinds of ceramics changes depending on factors such as temperature, ambient gas, and applied pressure. Such property changes can be used as sensor materials. In addition to the microscopic crystal structure, some electromagnetic functions can be modified by changing the macroscopic structures such as micrometer order pores and layers.

Keywords Dielectric materials · Magnetic materials · Semiconductor · Ion conductor · Superconductor

8.1 Classification of Ceramics with Electromagnetic Properties

A ceramic is a nonmetallic inorganic material. Different from metal, ceramic has been known as an electrically insulating material and, then, has been used as insulator for power line, substrate for electronic devices, etc. Since a series of oxide ceramics showing high temperature superconductivity was discovered in 1986, much attraction has been focused on its electric conduction. Before this, however, electric conduction of ceramics had been practically used as gas sensors and thermistors. Among them, varistor (variable resistor) and PTCR (positive-temperature coefficient resistor) are based on the electro-resistance change of grain boundaries depending on the electric field and temperature. Such properties cannot be seen in single crystalline materials without grain boundaries.

A. Kishimoto (✉)
Graduate School of Natural Science and Technology, Okayama University, Okayama, Japan
e-mail: kishim-a@cc.okayama-u.ac.jp

J. Hojo (ed.), *Materials Chemistry of Ceramics*,
https://doi.org/10.1007/978-981-13-9935-0_8

Magnetism and dielectric properties of ceramics are derived from spin and polarization, respectively. Some kinds of magnetic and dielectric materials, called ferromagnetic and ferroelectric ones, are composed of spin or polarization ordered regions (ferromagnetic domain or ferroelectric domain) reflecting their properties. Both ferromagnetic and ferroelectric materials are used for memory devices by detecting the remanent magnetic or polarized region previously recorded with magnetic or electric fields.

In the dielectric materials, there exist two kinds of particular subgroups in which the polarization varies depending on the impact stress and temperature. The former is called piezoelectric material and the latter pyroelectric one widely used for gas ignitor and infrared sensor, respectively. The piezoelectric materials can convert mechanical movement into electric field and vice versa, which is practicably used as piezoelectric transformer similar to winding transformer based on the electromagnetic induction.

In some kinds of solid, constituent ions can transfer by the applied field leading to ion conduction. They are called solid ion conductor or solid-state electrolyte analogous with liquid electrolyte. By replacing the liquid electrolyte with the solid-state one, reliability of electrochemical devices such as fuel cell and secondary battery would be improved.

As described above ceramics or inorganic polycrystalline materials possess wide variety of electric properties from insulator to superconductor. Such properties could be modified by the fabrication method and/or small amount of additives, which further varies depending on the temperature, ambient gas, or applied electric field. In this chapter, we introduce the origin of such attractive properties and describe the modification methods proposed or confirmed to improve them.

8.2 Fundamental Bases of Electromagnetism

In order to understand the macroscopic properties of ceramics, it is important to think the dependency of current density (j) or electric flux density (D) on the applied electric field (E) and dependency of magnetic flux density (B) on the applied magnetic field (H). The coefficients for these relations are called conductivity (σ), dielectric properties (ε), and permeability (μ). The electric flux density and magnetic flux one would increase when passing through some dielectric or magnetic materials, which can be described with ratio or increment compared with the vacuum case, as follows:

$$D = \varepsilon E = \varepsilon_r \varepsilon_0 E = \varepsilon_0 E + P \tag{8.1}$$

$$B = \mu H = \mu_r \mu_0 H = \mu_0 (H + M) \tag{8.2}$$

where ε_r is defined as relative dielectric constant, ε_0 as dielectric constant of vacuum, P as polarization, μ_r as relative permeability, μ_0 as permeability of vacuum, and M as magnetization. From the latter two equations, we obtain

$$P = \varepsilon_0(\varepsilon_r - 1)E, M = (\mu_r - 1)H \tag{8.3}$$

In the following section, electric or magnetic field response of P or M will be discussed in dielectric or magnetic materials.

8.3 Dielectric and Magnetic Materials

8.3.1 Polarization and Magnetization

Dielectric properties are derived from the polarization and magnetic ones from magnetic moment mainly based on spin one. There are several scales of polarizations such as electronic, ionic, orientational, and interfacial ones. The dielectric polarization decreases with applied field frequency because the inversion of the polarization direction cannot follow the change of applied field when the polarization scale becomes large.

The magnetic moment of an atom can primitively be explained by two types of current associated with the revolution and rotation of the electrons. Since the former is usually cancelled by those of the other electrons, almost all the magnetic moment of an atom is derived from the latter or spin. Such spin-derived magnetism disappears by accommodating a couple of spins antiparallelly in an orbit. Magnetism of atoms also diminishes due to so-called superexchange interaction in atom-oxygen-atom arrangement.

The magnetic materials containing magnetic atoms can be divided into paramagnetic one and the other materials. In the former the magnetic moments orient randomly and in the latter align or can align in one line. The former material is called paramagnetic ones, and the latter are further divided into ferromagnetic, antiferromagnetic, and ferimagnetic ones. In the ferromagnetic materials, all the magnetic moments can be oriented into one direction, showing a macroscopic magnetization. In the antiferromagnetic materials, half of the magnetic moments orient in the opposite direction to cancel the magnetism with each other, resulting in no apparent magnetization. In ferrimagnetic materials, however, the oppositely directed magnetic moments do not exactly cancel, so that a net magnetic moment results. The magnetization still existing after removing the magnetic field is called spontaneous magnetization.

8.3.2 Ferroelectric and Ferromagnetic Domains

Like ferromagnetic materials dielectric materials with oriented polarizations are called ferroelectric ones. In ferroelectric materials there exists spontaneous polarization like spontaneous magnetization in ferromagnetic ones. The spontaneous magnetization and polarization usually are oriented within microscopic regions in ceramics grains which are called magnetic and ferroelectric domains, respectively.

8.3.3 Domain Movement and Hysteresis

Just after the fabrication of both ferroelectric and ferromagnetic materials, the oriented directions within the domains are randomly arranged mutually indicating that net polarization or magnetization becomes zero. By applying the electric or magnetic field, domain walls move to form large domains in a ceramic grain coming to existence with macroscopic polarization or magnetization.

The above domain movement would be explained for ferromagnetic materials in detail by using Fig. 8.1 as follows. Before the application of magnetic field *H*, magnetization (*M*) equals zero as mentioned before. (1) As the material is exposed to an increasing magnetic field (*H*), apparent *M* becomes large due to the domain wall movement to enlarge the domain area of which the direction of magnetic moment accords with that of *H*. (2) When the domain wall moves to the grain boundary, the grain is composed of only one domain. At this time, however, the magnetic moment of the grain is not necessary to be parallel to the *H*. (3) As the field continues to

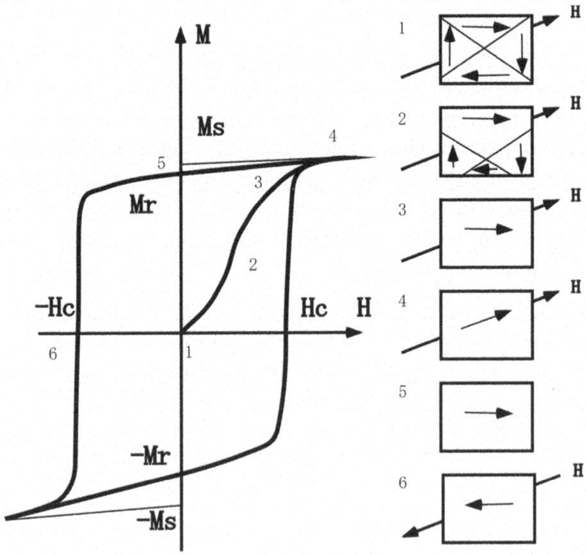

Fig. 8.1 M-H hysteresis of ferromagnetic materials and evolution of magnetic domains

increase, magnetic domains within the grains rotate to align with it. At this point the material becomes saturated or any further increase in *H* will not increase *M*. (4) This value at the maximum field is named saturation magnetization *Ms*.

Reducing the magnetic field from this value does not cause a drastic reduction in magnetization only based on the relaxation of the domain rotation. When the external field has been reduced to zero, there remains a significant remanent magnetization *Mr*. (5) Such magnetization remains when the applying magnetic field is changed to opposite sense. When the magnetic field in the opposite sense exceeds a critical value, the magnetic domain switches to align themselves with the field. This critical field is called coercive magnetic field *Hc*. (6) Increasing the magnetic field further in the negative direction causes the domains to grow until the material is once again saturated with -*Ms*. Reversing the field causes the magnetic properties to draw a square in *M-H* graph. In such materials two different *M* values would be obtained for *H* value. In other words, a property depends on the previously experienced field in addition to the field at that time. Such profile is called hysteresis behavior.

8.3.4 Electromagnetic Memory

In the previous section, properties of ferromagnetic materials under magnetic field have been described. Properties of ferroelectric ones can be similarly described by replacing magnetic moment *M* for polarization *P* under electric field *E* where *P* aligns within the ferroelectric domain, (in the latter case, remanent polarization *Pr* and coercive electric field *Ec* correspond to the *Mr* and *Hc* in ferromagnetic materials, respectively).

In both ferromagnetic and ferroelectric materials, magnetic moment and polarization remains after removing the field, and then, both materials have been used for electromagnetic memory devices. Ferromagnetic memory devices are preceded in a practical use. Among them, ferrites are predominantly used. In order to improve the recording density, the magnetized region should be decreased in addition to secure a considerable *Hc* value so as not to be erased by the ambient magnetic field. However, if the *Hc* becomes larger than the magnetic head performance, recorded data cannot be overwritten. Also the larger *Mr* value is desirable for reading the recorded data. Recently, magnetoresistance effect is tried to be used for high-sensitive reading head feasible to detect a weak magnetization of a narrow region.

Ferroelectric random-access memory (FeRAM) is a nonvolatile recording medium different from dynamic random-access memory (DRAM) and static random-access memory (SRAM). As nonvolatile recording memory, flash memory and electrically erasable and programmable read-only memory have been practically used. FeRAM is superior to these devices in read-write speed and operation power needed. Three kinds of perovskite oxide, $Pb(Zr,Ti)O_3$, $SrBi_2TaO_9$, and $BiLaTiO_3$, are mainly examined to use as FeRAM material from the view point of durability for repetitional read-write cycle in addition to large *Pr*.

8.3.5 High-Frequency Material and the Rader Absorbent Material

In the previous section, external magnetic or electric field varied slowly enough to follow the inversion in magnetic moment or polarization. If the field varying speed or frequency is increased, the components capable to follow the direction change decreases, leading to smaller B or D with the same external field. In other words, the coefficients connecting these values with the external fields, i.e., μ and ε, become small. In addition, phase delay (δ) between these value and external fields becomes predominant at higher frequency.

As described above there are frequency dispersions in permeability and dielectric constant, and these values are usually represented by complex numbers composed of real part and imaginary part. Complex permeability ($\mu*$) and complex dielectric constant ($\varepsilon*$) can be expressed as follows:

$$\mu^{*} = \mu' - j\mu'' \tag{8.4}$$

$$\varepsilon^{*} = \varepsilon' - j\varepsilon'' \tag{8.5}$$

where $'$ and $''$ represent real and imaginary parts, respectively. The tangent value of deviation angle (δ) from the reality relates to the energy dissipation, then, which is defined as dissipation factor (tan δ) represented by the ratio of real and imaginary parts (tan $\delta = \mu''/\mu'$ and tan $\delta = \varepsilon''/\varepsilon'$).

Contrary to this situation, high energy loss at high frequency in magnetic or dielectric materials is tied to be used for the Rader absorbent materials. Recently, electromagnetic wave with frequency range around GHz has come to use vigorously in cellular phone and wireless LAN, etc. Electromagnetic interferences between electromagnetic wave and electric circuit would lead misfunction and transmission error. Under these circumstances, two distinct functions are required: these are not to emit unnecessary electromagnetic wave and not to be affected by the outer electromagnetic wave. To accomplish this electromagnetic compatibility, much attention has been paid on the Rader absorbent properties rather than conventional electromagnetic shield. In order to design a favorable Rader absorbent device, controlling the high-frequency dielectric constant and permittivity is indispensable. For this purpose, silicon carbide and ferrite-based materials are usually used by modified those properties.

8.3.6 Ceramic Filter

In the previous section, absorption properties in the dielectric and magnetic materials have been described; however, an electromagnetic wave does not necessarily be absorbed in solid. One of the most important components of microwave telecommunication is receiver. As for receiver devices, dielectric ceramics are widely used as a key element, which is called dielectric resonator because it resonates with specific length of telecommunication waves. At that resonance length of wave, the specific impedance of the input side becomes the maximum to minimize the electric current. As a result the target telecommunication wave transmits the resonator without decay, while the other length of electromagnetic waves decays substantially. In this manner, dielectric resonator can transmit a specific length of electromagnetic wave selectively; it is also called a ceramic filter.

Dielectric ceramics with relatively large dielectric constant work as a resonator because a standing wave is formed so as to confine the energy of electromagnetic wave by reflecting the wave at the boundary between the dielectric material and free space. In a dielectric material, the wave length of any electromagnetic wave decreases in proportion to $1/\varepsilon_r^{1/2}$. As a result, a resonator can be miniaturized in size by using dielectric materials compared with the hollow resonator [1].

In recent mobile communication, microwave with frequency range from 800 MHz to 2.4 GHz is usually used. The correspondence wave lengths are in the order of 10 cm which can be reduced by using dielectric materials, contributing to miniaturizing the ceramics filter. For further reduction in the size of cellular phone and PHS, filter dielectric materials with smaller dielectric loss and smaller temperature dependence in addition to high dielectric constant have been explored.

8.3.7 Multilayer Capacitor

Ceramic capacitor is a simple but the most important application of dielectric ceramics. The capacitance C of a ceramics capacitor with electrode area S and electrode distance t can be expressed as follows:

$$C = \frac{\varepsilon_r \varepsilon_0 S}{t} \tag{8.6}$$

In order to obtain a larger capacitance, macroscopic structures are designed so as to enlarge S and to reduce t, in addition to employ a dielectric material with larger ε_r. Recently, high-capacitance ceramic capacitor takes a layered structure, in which multilayered electrodes inserted in a high ε_r dielectric ceramics connect side electrode alternately. With increasing the electrode number, in other words, parallel connected electrode contributes to enlarge S at the same time with decreasing t.

As dielectric materials with high dielectric constant, lead-based dielectric materials called relaxer or barium titanate-based materials are used. Characteristic phase transition called diffused transition is used in the former. In the latter materials, various additives are used to change the temperature dependence of dielectric constant. For example, additives to decrease the phase transition temperature are called shifter and to reduce the temperature dependence of dielectric constant are called depressor.

8.3.8 Piezoelectric and Pyroelectric Materials Based on the Symmetry of Their Crystal Structure

In Sect. 8.2, the dielectric constant was defined as the coefficient connecting electric flux density (D) with electric field (E). In reality, D is also dependent on stress (T), temperature change (dQ), and magnetic field (H). Assuming that the effect of magnetic field can be negligible, D can be expressed as an exact differential formula as follows:

$$D = (\partial D/\partial E)E + (\partial D/\partial T)T + (\partial D/\partial dQ)\mathrm{d}Q \tag{8.7}$$

The coefficients for the latter two intensive valuables are represented as $(\partial D/\partial T) = d$ named as piezoelectric constant and as $(\partial D/\partial \mathrm{d}Q) = p$ named as pyroelectric constant, or stress and temperature dependencies of electric flux density.

The elements of symmetry are utilized to define symmetry about a point in space. All crystals can be divided into 32 different point groups. Of the 32 point groups, 21 classes do not possess a center of symmetry. One out of the 21 classes is not piezoelectric because of other combined symmetry elements. As a result 20 point groups possess piezoelectricity. The piezoelectricity is defined as electric polarization produced by mechanical stress. In the same material, converse or inverse effect can be seen. The converse effect is associated with the mechanical movement generated by the application of an electrical field.

The first and second terms of the previous equation and the relation to give elastic strain S are called piezoelectric fundamental formula.

$$D = \varepsilon^T E + dT \tag{8.8}$$

$$S = dE + s^E T \tag{8.9}$$

where s^E is called compliance which equals $(\partial S/\partial T)$ or coefficient between strain and stress. The ε^T represents the dielectric constant under constant T. Since the strain and the electric flux density can be expressed by using the Gibbs free energy as $S = (\partial G/\partial T)_{\mathrm{E}}$ and $D = - (\partial G/\partial E)_{\mathrm{T}}$, the differential coefficient of the latter equation can be deformed as follows:

$$(\partial S/\partial E)_{\mathrm{T}} = -\left(\partial^2 G/\partial E\partial T\right) = -\left(\partial^2 G/\partial T\partial E\right) = (\partial D/\partial T)_{\mathrm{E}} = d$$

As a result, the second term of Eq. (8.8) and the first term of Eq. (8.9) have the same coefficient [2]. Since E and D are vector (the first-order tensor) and T and S are the second-order tensor, ε^T, d, and s^E are the second-, third-, and fourth-order tensor, respectively.

There are 10 crystal classes out of above 20 ones possessing spontaneous polarization in which symmetry center of cation does not accord with the anion one. This polarization also changes with temperature; hence, it is called

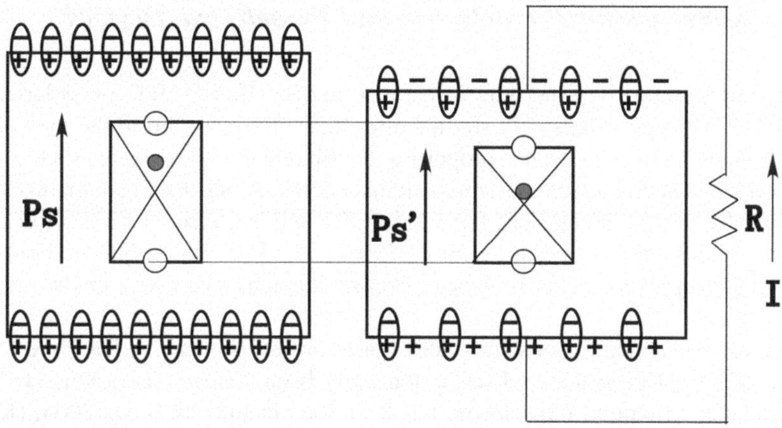

Fig. 8.2 IR detection mechanism of a pyroelectric material (left, room temperature; right, on IR irradiation)

Fig. 8.3 Interrelationship of piezoelectrics and pyroelectrics based on the crystal symmetry

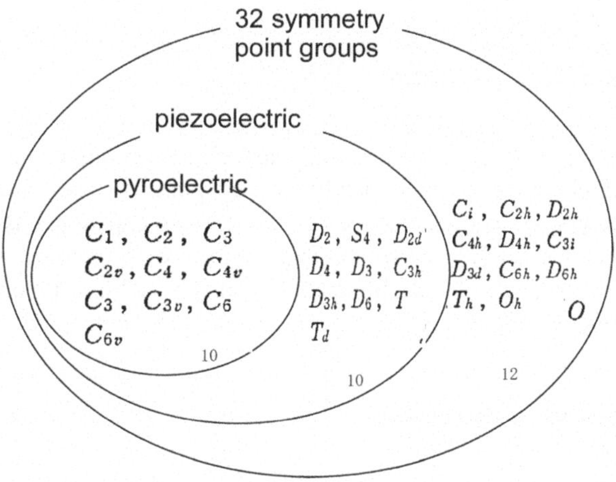

pyroelectricity. The temperature coefficient of polarization p is usually negative because the polarization decreases with temperature.

In a pyroelectric material, the spontaneous polarization is cancelled by a kind of floating charge at a given temperature. On a sudden temperature increase, the floating charge is released due to the polarization decrease. The charge movement can be detected as a current of outer circuit giving an infrared sensing device (Fig. 8.2). Such sensor does not need standby power, and then is now widely used as intrusion alarm and thermal imaging.

The ten spontaneous polarized materials have a special category called ferroelectricity, in which the polarization is reversible by an electric field of some magnitude. It can be concluded that the ferroelectric materials are sufficiently pyroelectric one and piezoelectric one. This relationship is illustrated in Fig. 8.3.

8.3.9 Materials for Piezoelectric and Pyroelectric Devices

Piezoelectricity and pyroelectriciy in barium titanate ($BaTiO_3$:BT) on polarization treatment have been extensively studied after their discovery. From the viewpoints of piezoelectric and pyroelectric properties, lead-based complex oxides such as lead titinate ($PbTiO_3$:PT) and lead zirconate titanate ($Pb(Zr,Ti)O_3$:PZT) are mainly used. The latter has a solid solution of ferroelectric PT and antiferroelectric $PbZrO_3$. The latter is usually used with composition of Zr/Ti = 53/47 (morphotropic boundary phase) at which piezoelectric properties become the maximum with flexible crystal structure.

Recent environmental conscious does not afford to use lead, so lead-free piezoelectric or pyroelectric material has extensively been studied. $(Bi_{1/2}Na_{1/2})TiO_3$ or bismuth layer-structured ferroelectric has been long examined. Since 2004, (K,Na) NbO_3-based ceramics have attracted attention because its piezoelectric properties are comparable to those of PZT, while neither lead nor bismuth is contained.

8.3.10 Actuator

When voltage is applied on a piezoelectric ceramics, it will expand or contract in the thickness direction depending on the polarity of the voltage with respect to the poled direction. The magnitude of these movements depends on both voltage and d value but is usually very small. Then microscopic ceramics bars are assembled to obtain larger displacement. In general layer type or bimorph one is usually used as shown in Fig. 8.4.

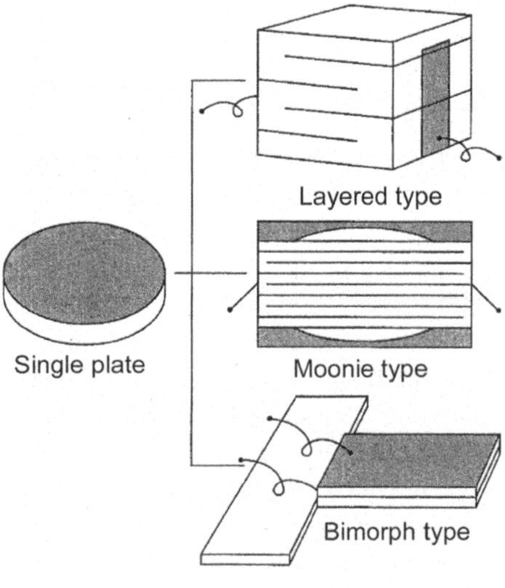

Fig. 8.4 Variety of design in ceramics actuators [3]

In the former layer type, about a hundred ceramics layers were piled. Electrodes were formed between each layer to make interdigital ones. As a result, power force as large as 100 kgf can be obtained with relatively small applied voltage around 100 V. Another merit of this type is quick response (10 μs). On the other hand, it has a drawback of small output displacement as large as 10 μm because it is only the sum of each layer strain.

Another type of actuator bimorph is composed of two rectangular plates with the same dimension but oppositely poled. The bender deflects by applying the voltage because one expands and the other contracts. It can easily create bending displacement over 100 μm; however, its response time is rather late (1 ms), and its available force is inevitably small.

Relatively novel type of actuator called *moonie* has attracted attentions because its displacement is as large as that of layer type, and its generated power and response time are better than those of bimorph type. As shown in Fig. 8.4, crescent-shaped metal plates are attached to top and bottom sides of the layered actuator to make bridges. In addition to the vertical displacement of layered actuator, horizontal contract is transformed to the vertical displacement with the metal plates.

8.3.11 Piezoelectric Transformer

In this section the analogy of magnetic and dielectric materials has been described. In the following section, piezoelectric materials which are similarly used for voltage converter like conventional winding transformer will be described.

Conventional voltage converter based on the electromagnetic induction law is called winding transformer in which input/output voltage ratio is proportional to the ratio of turns number. A magnetic material with high permeability is used to transmit magnetic flux from the primary to the secondary side, which is called transformer core. A piezoelectric transformer can convert voltage by using electromechanical interaction of piezoelectric material.

As shown in Fig. 8.5, two differently poled piezoelectric materials are joined to make a piezoelectric transformer which is called the Rosen-type one. When a voltage with a resonant frequency is applied on the primary side of piezoelectric material poled vertically, strain generates also vertically, and at the same time horizontal

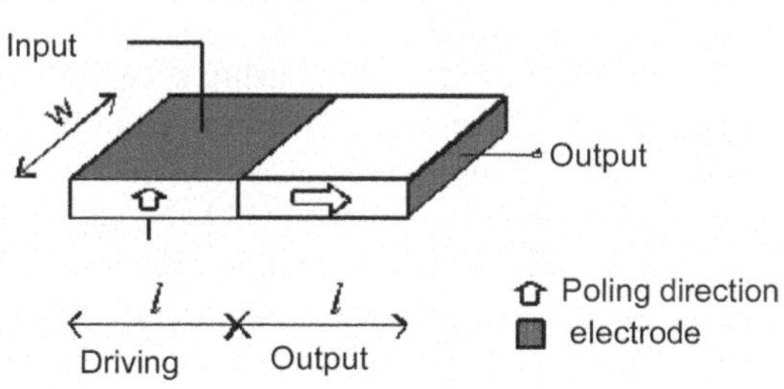

Fig. 8.5 Schematic illustration of piezoelectric transformer (Rosen type)

strain also generates. That horizontal strain is transmitted to the horizontally poled secondary side to output a voltage by the piezoelectric conversion.

Boosting ratio of piezoelectric transformer is theoretically large, and it does not generate magnetic field. On such reason, piezoelectric transformer is widely used for voltage converter in backlight for liquid crystal display.

8.4 Semiconductor Ceramics

In a dielectric material, charged particles like ion, electron and hole cannot move through the solid but only sift from their stable positions, then it is called electrically insulating material. Contrary to this, there are a series of materials called conductor or semiconductor in which electron or hole moves continuously.

As an index of electric conduction, conductivity σ or the coefficient connecting current density j and electric field E is usually used similarly to the cases of dielectric or magnetic materials. In detail, the resistance R is proportional to its length l and inversely proportional to the sectional area S, which relationship can be expressed as follows:

$$R = \rho \frac{l}{S} \tag{8.10}$$

where the coefficient ρ is called resistivity and the reciprocal of resistivity is defined as conductivity σ. Solid materials are classified by σ, i.e., materials with σ larger than 10^5 Sm are called conductor (metallic conductor); those smaller than 10^{-6} are called dielectrics. The solids with conductivity between these values are called semiconductors.

8.4.1 Electric Orbit and Band Structure

The conductivity of a material can be estimated by the band structure at which electrons are allowed to exist. Such energy band is constituted of electron orbit of individual atoms. In other words, electron orbit of constituent atoms determines the energy band of a solid.

Figure 8.6 shows the relationship between molecular orbital and the band structure composed of two kinds of atoms A and B. If the difference of the atomic energy levels between A and B is small, difference between bonding orbital and antibonding one becomes small leading to small band gap in solid composed of A and B (smaller case). Band build from the bonding orbital is called valence band and that from the antibonding one conduction band. On the other hand, if difference in the atomic energy levels is large, difference in energy level between bonding orbital and antibonding one becomes large, resulting in wide band gap (larger case).

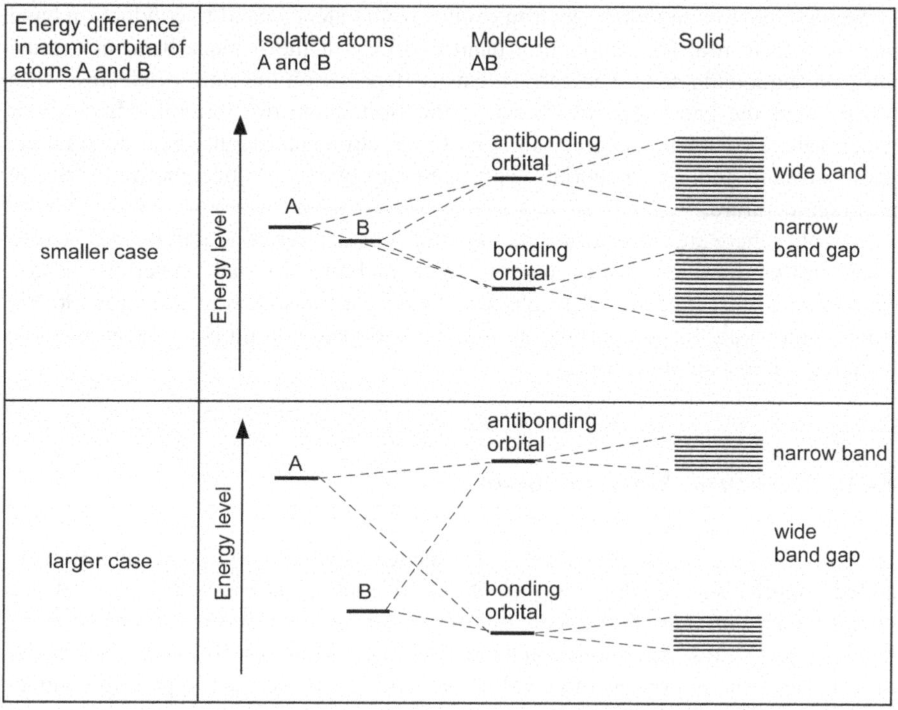

Fig. 8.6 Energy levels allowed to take for an electron in isolated atoms, molecule, and solid [4]

Band gap is also called forbidden band at which electrons are not allowed to exist. Although the fully occupied electrons at an energy band cannot move, partially occupied electron in a band can move to transport electric charge.

When the electrons are put into the valence band and the last electron exists at the middle of the valence band, it can move freely through the solid via the unoccupied valance band. In the other solid, valence band is fully occupied, and the conduction band is empty. If the band gap is small, electrons at the valence band can be excited to conduction band with energy gain thermally or photonically, leading to electronic conduction. If the band gap is large, the electrons cannot reach to the conduction band remaining in an insulator.

After the electron excitation, a hole is remained in the valence band, and this hole has effectively positive charge to be mobile by bias application. Compared with the hole at the valence band, electron at the conduction band moves easily. It moves more easily when the conduction band is wide. The degree of movement of charged carrier is called mobility.

Conductivity is defined as electric charge passing through a unit area for 1 s by applying 1 V. Conductivity can be expressed as follows:

$$\sigma = en_e\mu_e + en_h\mu_h \tag{8.11}$$

where e is elementary charge, n_e and n_h are electron and hole densities, and μ_e and μ_h are electron and hole mobility, respectively.

Semiconductors in which electron excites from valence band to conduction band to make electron at the conduction band and hole at the valence band are called intrinsic semiconductors. The carrier density depends on the ratio of electron with energy over the band gap, which obeys the Boltzmann distribution. Materials of which valence band are occupied fully by the electron can roughly be judged by the width of band gap, i.e., materials with band gap over 2 eV become dielectrics at room temperature.

In reality there are some materials showing considerable conductivity even with a small carrier densities if they possess large mobility. In such materials electric carriers seldom scatter the inlet light, and then light transparency and large electric conduction would be compatible. Indium oxide doped with tin is a typical example realizing a transparent electrode.

8.4.2 *Extrinsic Semiconductor*

In addition to the direct excitation from valence to conduction band, intentionally added impurity state could supply electrons. If impurity atoms with larger valence than the parent ones are introduced, impurity states or donor levels are formed in the band gap just below the conduction band. Thermal excitation will lead to the transfer of electrons from donor level to conduction band, giving rise to n-type semiconductor. On the other hand, fewer valency impurities make impurity sates or acceptor levels in the band gap just above the valence band. It can accept the thermally excited electrons to give positive charged holes in valence band leading to p-type semiconductor. Extrinsic semiconductor is determined not by the width of band gap but by impurity levels from the band edges.

In a metal, its carrier density is insensitive to temperature; then the temperature dependency of conductivity is governed by the behavior of μ. In the case of semiconductor, on the other hand, conductivity increases with temperature due to the increased excited carrier with temperature elevation.

Figure 8.7 illustrates a temperature dependence of conductivity for n-type semiconductor. At lower temperature electrons are supplied mainly from the donor state of which excitation energy is small, giving a temperature dependency based on the energy gap between donor state and edge of conduction band. This temperature region is called impurity region. In the intermediate temperature, all impurity levels are effectively ionized, and the carrier density is temperature independent, which is called exhaust region. Finally at sufficiently high temperature, electrons would excite directly from valence band to conduction one, giving a temperature dependence based on the band gap. This is called intrinsic region.

By using above mentioned temperature dependence of resistivity, that is, conductivity, a device to estimate the temperature form the electric resistance is called thermistor. Commercially available thermistors are prevailing, of which resistance-temperature properties are controlled by mixing several transition metal oxides and thermal stability, or thermal response is adjusted by the configuration and cover coating.

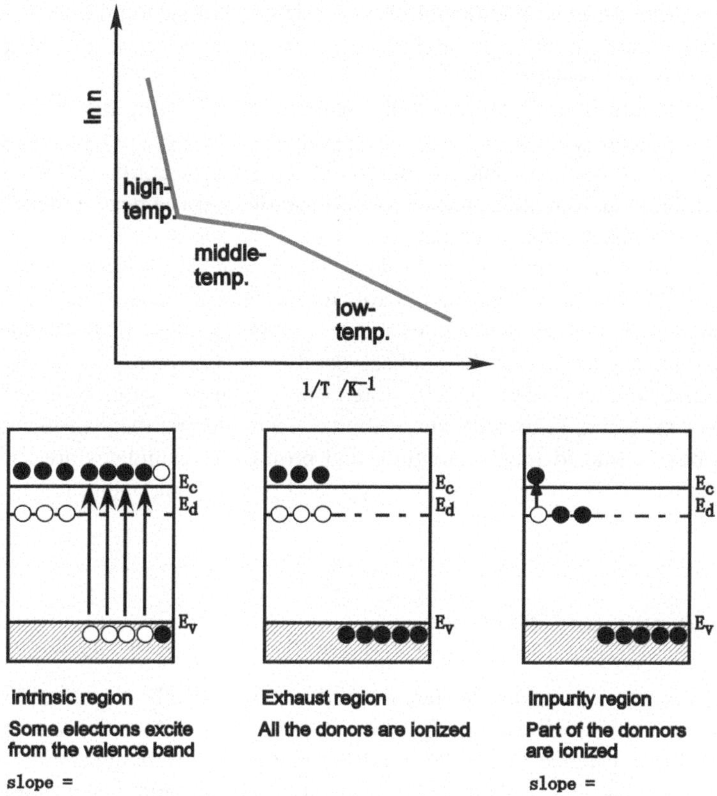

$$-\frac{E_c - E_d}{2k} = \frac{E_g}{2k} \qquad -\frac{E_c - E_d}{2k}$$

Fig. 8.7 Temperature dependence of carrier density in n-type semiconductors

8.4.3 Origin of Thermoelectric Power and Its Usage

In the previous section, the subjective solids are supposed to be heated uniformly; however, carrier density of high temperature side is large when the sample has temperature difference. In addition to the potential difference, concentration difference also acts as driving force, and the resultant electric current is called diffusion current.

When one of the edges of a semiconductor is heated, the carrier density of the heated side becomes large moving to the lower side by diffusion. If the transport carriers are electrons, lower-temperature edge gains negative charge generating a potential difference. This voltage is called thermoelectric power, and lower side potential divided by the temperature difference is defined as the Seebeck coefficient. When the transport carriers are electrons, the Seebeck coefficient becomes negative. In the case of hole, it becomes positive.

Since we can estimate a temperature as a reference of standard one, it is widely used as temperature monitoring device or thermocouple, which is advantageous for needing no electric power.

When p-n junction is fabricated by semiconductors with the large Seebeck coefficient and subjected to a certain temperature difference, practicably effective electric power can be obtainable leading to a thermoelectric converter. Such devices would be promising to realize an effective utilization of waste heat at thermal power generation and factory plants, so on.

In addition to the value of the Seebeck coefficient, low ohmic loss and large temperature difference are important factors for thermoelectric converter. As a result, it is usually evaluated by a figure of merit *Z* taking account of both electric and thermal conductivities beside the Seebeck coefficient. Practically available thermoelectric convertors with larger *Z* values are usually tellurite, antimonite, and silicide which have problems in toxicity and stability. To overcome these problems, oxide ceramics have attracted much attention, and promising candidates are ZnO-based compound and $NaCoO_2$ ceramics for n-type and p-type component, respectively.

8.4.4 Joining and Barrier

The band structure for electronic state in solid is schematically illustrated in Fig. 8.8 for semi-qualitative discussion. The lowest point of conduction band (*Ec*), the highest point of valence band (*Ev*), and the energy at the boundary of occupied and unoccupied carriers (E_F, Fermi level) are specially important energy levels. The corresponding energies as reference of vacuum level (*VL*) are called electron affinity, ionization potential, and work function.

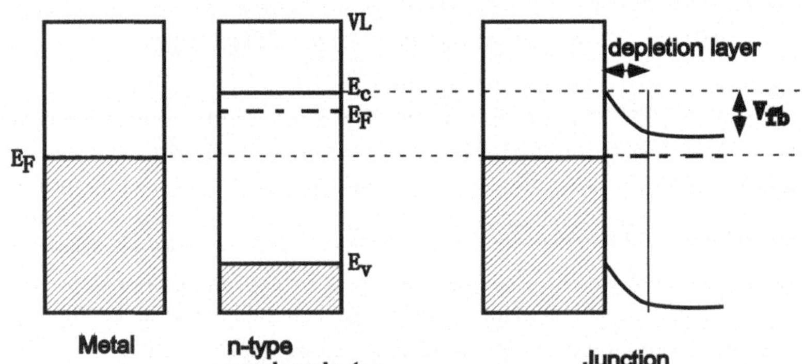

Fig. 8.8 Band structures of metal and n-type semiconductor. The Schottky barrier is formed by junction

When materials with different Fermi levels are contacted, electrons would transport from high-level material to low-level one. In the case of metal and n-type semiconductor, electrons transport form semiconductor to metal to make the Fermi levels coincide as shown in Fig. 8.8. An electron depletion layer or exhaustion layer is formed in the semiconductor side at the vicinity of the boundary. Positively charged donor ions are left behind in this region building an electrostatic potential gradient. Such a gradient becomes a barrier for electron transport, which is called the Schottky barrier. The height of built-in potential as reference of that far enough from the junction is called flat band potential (V_{fb}).

8.4.5 Rectification at Metal-Semiconductor Junction

When a bias voltage is applied across the boundary with semiconductor side going into positive, the Ec is relatively lowered with increasing the potential difference. At this time practically no electrons can climb the potential energy hill from metal to semiconductor. For back voltage when a bias negative voltage is increased to semiconductor, the Ec increases reducing the barrier height. The barrier height becomes zero when the applied bias equals V_{fb}; then, more electrons flow from semiconductor to metal.

The abovementioned relationship is illustrated in Fig. 8.9 at which forward bias means semiconductor side is negative and backward one means semiconductor side is positive. A large current will flow if we apply a voltage in one direction, but if the voltage is in the opposite direction, only a very small current will flow. Such property is called rectification which is also seen in p-n junction of semiconductors and used for current limiting device.

8.4.6 Built-in Barriers at the Surface and Grain Boundary

For a certain combination of semiconductor and metal or considerable percent of semiconductor junction, some degrees of depletion layers are formed possessing a high-resistivity boundary or surface (Fig. 8.10).

When the Schottky barrier is formed by using a specific metal electrode on a semiconductor, the rectification properties based on the junction might be misunderstood as an intrinsic property. To avoid such misevaluation, metal electrode with small work function should be used so as not to form the Schottky barrier.

There exist surface and grain-boundary levels based on absorption and impurity. If these levels are lower than the Fermi level of bulk or gain, the Schottky barriers form. When such barrier can be varied by some environmental feature, resistivity change can be utilized for a sensing device.

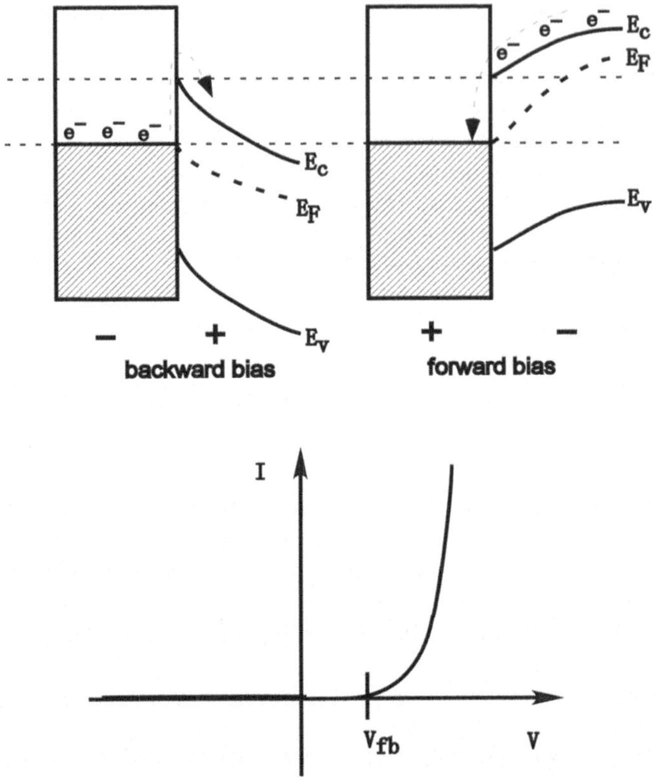

Fig. 8.9 Rectification of metal-semiconductor junction on bias application

8.4.7 Sensing Mechanism of Semiconductive Gas Sensor

Semiconductive gas sensors are usually used in ambient air. In such case adsorbed oxygen is charged to be negative due to its large electron affinity. Those electrons are supplied from the surface resulting in large surface resistivity. By using band structure, in other words, adsorption levels formed below the Fermi level capture the electrons at the vicinity of bulk surface to make a depletion layer.

When flammable gases such as hydrogen, carbon monoxide, and hydrocarbon are introduced to an oxygen adsorbed semiconductor, the introduced gases react with adsorbed oxygen giving water and carbon dioxide. The electrons used for negative adsorption return back to semiconductor bulk reducing the height of the depletion layer. Concomitant resistivity change can be utilized for flammable gas sensing. The ratio of resistivity change depends on the amount of reacted oxygen or flammable gas concentration when the amount of absorbed oxygen is sufficiently large. Since the sensitivity increases with increasing the surface to bulk ratio, thin film or porous morphology is usually used as sensor configuration. In addition to the sensitivity, response time and selectivity are also important, which are tried to be enhanced by surface modification.

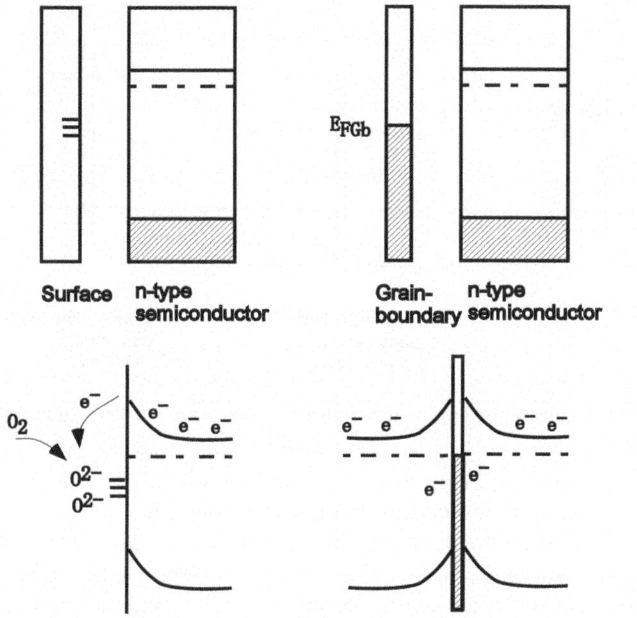

Fig. 8.10 Barrier formation in connection to the surface and grain boundary levels

8.4.8 PTCR Device and Varistor Based on Grain Boundary Barrier

Grain boundary impurity levels form symmetric barrier called the double Schottky one. The figure of this barrier can be calculated using the Poisson's equation giving the resistivity ρ:

$$\rho = \rho_0 \exp(\phi/kT) \tag{8.12}$$

where ρ_0 is the standard resistivity, k the Boltzmann's constant, T the absolute temperature, and ϕ the barrier height.

Since the height of barrier is inversely proportional to the relative dielectric constant, electric resistance varies with dielectric constants. Barium titanate has tetragonal-to-cubic phase transition at around 120 °C, below which it possesses spontaneous polarization showing ferroelectricity. Its permittivity shows peak at this temperature decreasing suddenly with temperature. When impurities with valency larger than titanium and lanthanum are doped into barium titanate, it becomes n-type semiconductor. The resistivity of doped barium titanate changes non-monotonically. This property is called PTCR (positive temperature coefficient of resistivity) which is widely used as overheating protection device for dryer and cooking heater. There are several models to explain the PTCR effect of barium titanate. Among them the

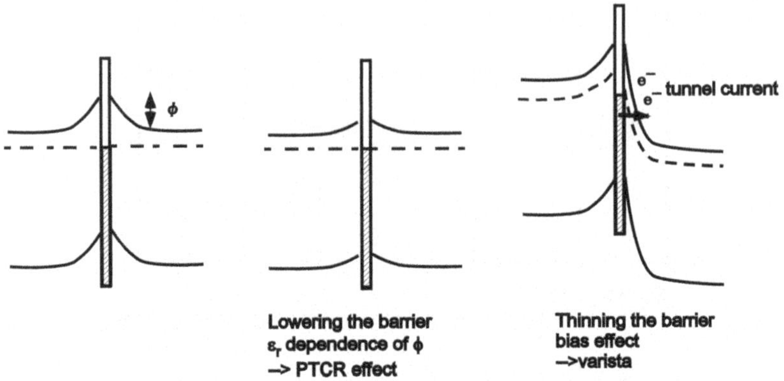

Fig. 8.11 Barrier height depending on the temperature and tunnel current under large bias

most probable one is the Heywang model in which height of the double Schottky barrier is correlated with the change in dielectric constant [5].

Another device using the double Schottky barrier at the grain boundary is varistor or abbreviation of variable resistor. This material can be regarded as insulator up to a certain electric field or breakdown field, above which it becomes conductor. When a varistor is directly connected to a sensitive device, it can be protected from transient overvoltage. In addition to the protection of electronic equipment, a major application of varistor is for the protection of electronic power distribution and transmission system which require reliability with voltage up to a megavolt range.

The usual varistor is a zinc oxide-based ceramic essential to be doped with bismuth oxide or praseodymium oxide. These dopants are thought to form grain boundary levels building the double Schottky barrier. As shown in Fig. 8.11, with low applied voltage, both the height and the width of the depletion layer are sufficiently large resulting in few conducting electron. When the voltage exceeds some critical value, larger number of electrons can conduct by a tunneling process through thinned depletion layer.

8.5 Ion Conductor

Electric charge transport carried by ion is called ionic conduction. An electrolytic solution such as diluted sulfuric acid or molten salt exhibits conspicuous ionic conduction. Even with a solid state, some materials show sufficiently large ion conduction comparable to liquid electrolyte, which is called solid electrolyte.

Conductive solid can be classified by the ratio of ionic conduction to the total conduction. If almost all the conduction is attained by ionic one, such material is named ionic conductor. When the contribution of electron and ions to the total conductivity is comparable, they are called mixed conductor. In general solid electrolyte means materials with large ionic conductivity; mixed conductor of which contribution of ionic one is small may be sometimes called solid electrolyte.

Compared with the electron, ion has large mass and volume. Among them ions with smaller radius and valence can move relatively easily. Since the ionic radii of fluoride and oxide ions are large, higher temperature is needed to show large ionic conduction of these ions.

8.5.1 Crystal Structure of Ionic Conductor

The major structural feature of ionic conductor could be divided into the following four categories: (1) partially occupied sublattice sites, (2) defect structure, (3) layer or network structure, and (4) amorphous.

In the partially occupied sublattice sites structure, the number of availably occupied sites is larger than that of real existing ion. As a result, mobile ions are distributed statistically on these sublattice sites. The crystal structure of this typical material or α-AgI is illustrated in Fig. 8.12. This structure is based on body-centered cubic iodide anion with silver cation in the possible sites. The silver cations exist mainly at the tetrahedron sites made of iodide anions. In addition to these 12 four coordinate sites existing in the unit cell, there are 6 six coordinate and 24 three coordinate sites. These three sites are energetically equivalent; two silver cations can move through these 42 sites.

Below 147 °C α-AgI transforms to β-structure suddenly decreasing the ionic conductivity. In order to overcome this drawback, high-temperature phase was tied

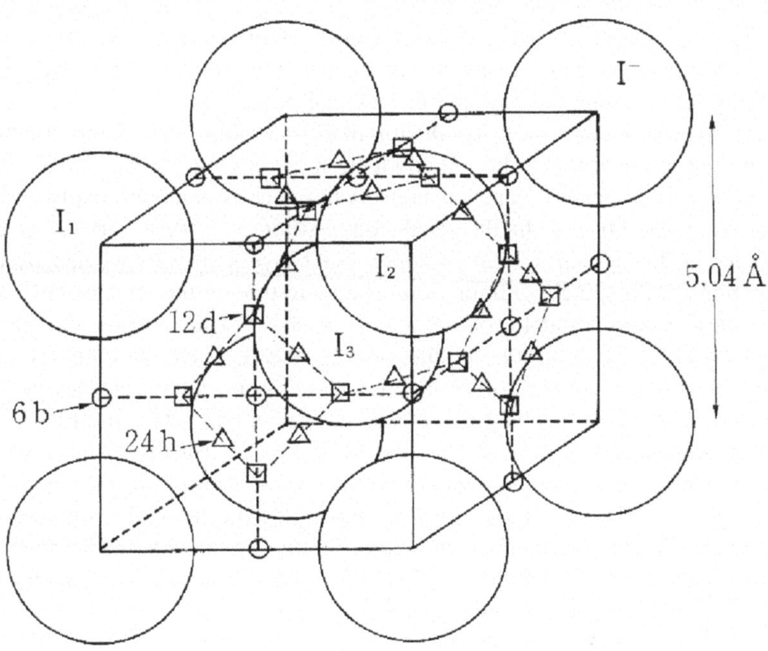

Fig. 8.12 Crystal structure of α-AgI. I sites; large spheres, Ag sites; the other symbols [6]

to be stabilized by vitrification. To realize the partially occupied structure at room temperature, some derivatives such as $RbAg_4I_5$ and $Rb_4Cu_{16}I_7Cl_{13}$ have been synthesized.

Another type of ionic conductor has defect structure in which ion can move through the defects formed by the addition of impurities. Stabilized zirconia, in which cubic high- temperature fluoride phase is maintained by adding CaO or Y_2O_3, is denoted as CSZ and YSZ. At this time low-valence cations are substituted by the zirconium ion; oxygen vacancy equivalent amount of the dopant would form as follows:

$$2(1 - x)ZrO_2 + xY_2O_3 \rightarrow Zr_{2-2x}Y_{2x}O_{4-x} \quad (8.13)$$

$$(1 - y)ZrO_2 + yCaO \rightarrow Zr_{1-y}Ca_yO_{2-y} \quad (8.14)$$

These vacancies exchange with the lattice oxide ions resulting in ion conduction.

For practical application of oxide ion conductor, it is also important that the ionic conduction is predominant for a wide range of oxygen partial pressure. The oxide ion conductivity of ceria or bismuth oxide is larger than YSZ; however, they are readily reduced accompanied by electronic conduction. Taking into account the atmosphere exhibiting pure ionic conduction, stabilized zirconia is mainly used as oxide ion conductor.

In addition to yttria and calcia, stabilization of zirconia is realized by a variety of rare earth oxides. All systems exhibit the maximum in the ionic conductivity at specific composition depending on the doped cations. The decrease in conduction with increasing the defect over the maximum composition could be explained by an association of charged defect and oxide ions. Although Sc_2O_3 or Yb_2O_3 doped zirconia shows larger ionic conductivity than that of YSZ, YSZ is still mainly used taking into account the cost and mechanical properties.

In the layer-structured ionic conductor like β-alumina, large ionic conduction would realize in plane when the ion conduction paths connect two dimensionally. The β-alumina is a generic name of nonstoichiometric compound expressed by a formula $Na_{1+x}Al_{11}O_{17+x/2}$. In the ideal composition $x = 0$, its crystal structure belongs to the hexagonal system in which NaO layers are sandwiched by dense $Al_{11}O_{17}$ block with spinel configuration. There are three kinds of sites (BR, a-BR, mO) which can accommodate sodium ions (Fig. 8.13). They become energetically identical leading to sodium ion conduction like α-AgI. Since the ionic movement occurs in NaO plane, sodium ion conduction shows conspicuous anisotropy, or the conductivity vertical to the NaO plane becomes extremely small. In order to overcome this drawback NASICO ($NNa_2Zr_2P_2SiO_{12}$) and LISICON ($Li_{14}Zn(GeO_4)_4$) have been synthesized as compound with a three-dimensional conduction path.

Although amorphous material does not have periodic network structure, it has been attempted to be used as ionic conductor through its open structure. Compared with crystal glass, it is advantageous of which composition can be changed continuously and which can be processed easily.

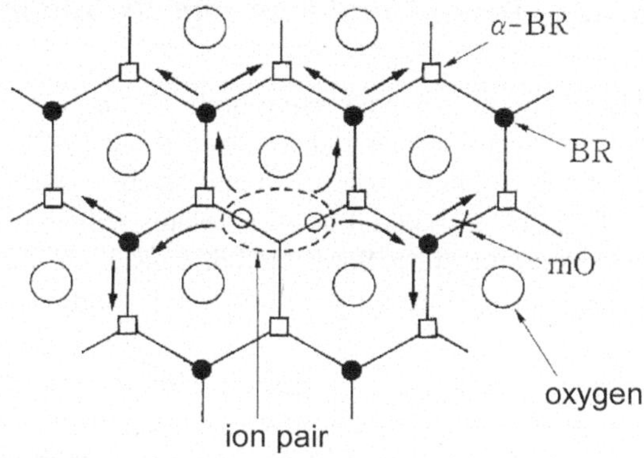

Fig. 8.13 Sodium conduction plane and conduction path of β-alumina [7]

8.5.2 Application of Ionic Conductor

The reliability of electrochemical devices would be improved by replacing the liquid electrolyte with solid one. Solid electrolyte would realize the high-temperature electrochemical devices at which liquid electrolyte cannot be used.

When a YSZ is used as a bulkhead and gases with different oxygen partial pressure ($P'o_2$, Po_2) are introduced into the two separated spaces, oxygen concentration cell would be composed with electromotive force E as follows:

$$E = \frac{RT}{4F} \ln \frac{P'_{O_2}}{P_{O_2}} \tag{8.15}$$

where R is the gas constant and F the faraday constant. If one of the partial pressure is already known, we can calculate the other one by measuring E and T.

This type of oxygen sensor has enabled to detect the oxygen solute in melting steel as high as 1900 K. Almost all of today's automobile is equipped with this type of sensor as air-fuel ratio sensor (λ sensor) to control the internal combustion engine. On the contrary, by applying a voltage over the electromotive force, it can be used as oxygen pump to carry oxygen from low to high partial pressure site.

If hydrogen gas is introduced to one site, reactions $H_2+O^{2-} \rightarrow H_2O+2e^-$ and $O_2+4e^- \rightarrow 2O^{2-}$ occur at hydrogen and oxygen sites, respectively. Electromotive force corresponding to ΔG for these reactions is generated to use as fuel cell. There are several types of fuel cells classified by the operating temperature. High-temperature fuel cell in which ceramics electrolyte is used has advantage in conversion efficiency and practicably used for on-site power generator.

8.5.3 Mixed Conductor and Its Application

Mixed conductors possessing both ion and electron conduction attract much attention as electrode material for several electrochemical devices. This is because the electrode is a place where ion and electron react in electrochemical reaction.

Cell active material accommodates reducing agent M as M^+, and it also needs electron conduction. Some kind of transient metal oxide T_mO_n with the highest metal valence generates easily following the reaction

$$T_mO_n + xM^+ + xe^- \rightarrow M_xT_mO_n \tag{8.16}$$

with large ΔG value. Manganese oxide (MnO_2) is a good example to use for commercial battery where H^+ accommodates. Thus described ion insertion into an electrode compound is called intercalation.

With a combination of lithium anode active material and cathode one which can reversibly intercalate-deintercalate lithium ion, lithium secondary battery can be assembled. Some kinds of layer materials such as TiS_2 and Li_xCoO_2 are promising candidates for cathode material as high energy density battery because their lithium diffusion and electric conduction are both large.

Electrochemical supercapacitor attracts much attention because it possesses both large capacity and high power density. It uses intercalation in addition to the electric double layer capacitance usually used for electric field capacitor.

Tungsten oxide has a perovskite structure with vacant twelve coordinate sites which easily accommodate alkali metal ions. Since the valence band level of alkali metal is higher than the conduction level of tungsten oxide, inserted metals supply electrons to the conduction band leading to metal conduction.

In addition to the mixed conducting properties, tungsten oxide shows coloration change accompanied by the insertion and deinsertion of alkaline metal. In detail, if negative potential is applied on tungsten oxide submerged in lithium ion dissolved electrolyte, it turns blue due to the lithium intercalation. The color disappears with applied positive potential. Such mechanism has been applied to a display called electrochromic one. This system does not need backup power to maintain the coloration, which is advantageous for large-area display and light shielding window. Since its passive coloration system is easy on our eyes, it has been tried to apply for electronic book.

8.6 Superconductor

After the first discovery of high-temperature superconductor in 1986, oxide superconductor monopolized the attention of solid-state physics community. This phenomenon was observed first in 1911. Since then, it has been reported in more than 1000 materials including metal compound, metallic alloy, and organic materials. Since the high-temperature superconductor exhibits superconductivity above the boiling point of liquid nitrogen ($Tc > 77$ K), the application environment has changed drastically because it does not need liquid helium.

In the present section, solid-state physical feature, properties, and application of superconductor will be described.

8.6.1 Crystal Structure of High-Temperature Superconductor

All the high-temperature oxide superconductors have perovskite-based structure with Cu-O layers. Cuprate superconductors are classified into the following four categories by the constituent:

YBCO: $YBaCu_3O_{7-x}$, $Tc = 92$ K.
Bi system: $(Bi,Pb)_2Sr_2Ca_2Cu_3O_{10+x}$, $Tc = 110$ K, etc.
Tl system: $Tl_2Ba_2Cu_2O_{8+x}$, $Tc = 125$ K, etc.
Hg system: $HgBa_2Ca_2Cu_3O_{8+x}$, $Tc = 135$ K, etc.

Crystal structures of YBCO, Bi system, and La_2CuO_4 are illustrated in Fig. 8.14. In any structures characteristic structure is commonly seen as hutched in Fig. 8.14. This is called CuO_2 layer or superconductive layer at which superconduction occurs when

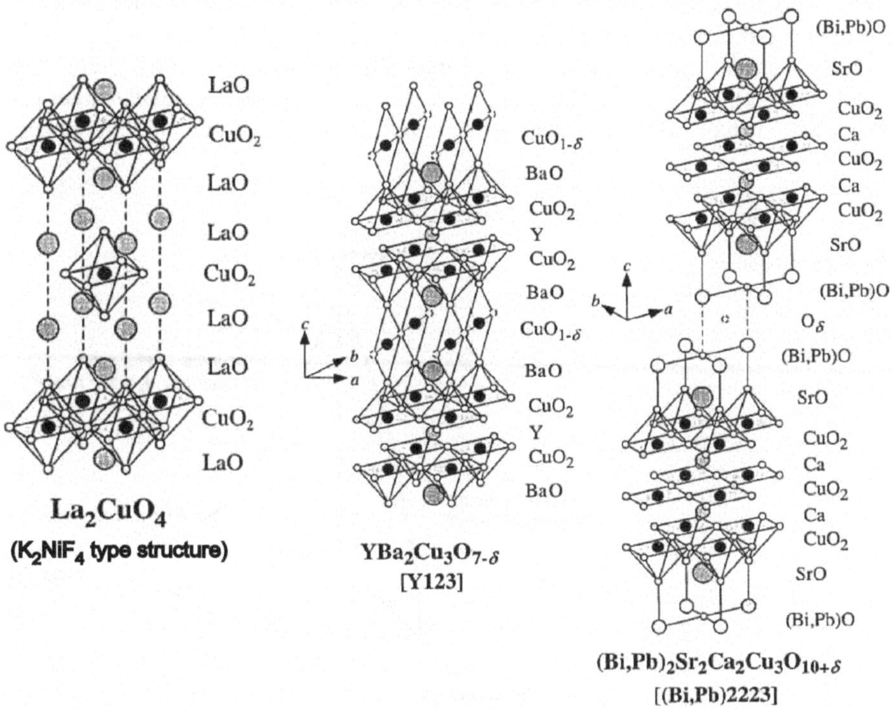

Fig. 8.14 Crystal structure of oxide superconductor [8]

introduced by electron holes. There are variety of structures between the CuO_2 layer, which are called blocking layer. In addition to inhibiting the connection of superconducting layer, it also supplies carriers to the superconducting layer then named as carrier supplying layer.

It is accepted that the superconducting cuprous oxide shows the maximum *Tc* when carriers are supplied with 0.15 electron unit per cupper ion. The *Tc* decreases when the carrier concentration deviates from the optimized one. Deficient or excess carrier concentration states are called under-dope and over-dope ones, respectively. Carrier concentration can also be controlled by a partial substitution with different valence ion, oxygen defect, and interstitial oxygen.

The electronic charge carriers in the high-temperature superconductor can be described fundamentally by the BCS theory proposed in 1957, which explains the unusual behavior as being due to electrons combining into Cooper pairs. The figure of the probability density of electron pairs is thought to be anisotropic like *d*-orbit.

8.6.2 Superconducting Properties

There exist three kinds of superconductivity regarding critical conditions as critical temperature *Tc*, critical magnetic field *Hc*, and critical current density *Jc*. As shown in Fig. 8.15, these three parameters correlate each other, and superconduction occurs within the *T-H-J* critical plane.

Critical current density could be theoretically increased up to 10^8 Acm^{-2}, at which electron pair breaks. The realistic *Jc* is one to four orders smaller than this theoretical value. The degradation can be ascribed to the inhomogeneous

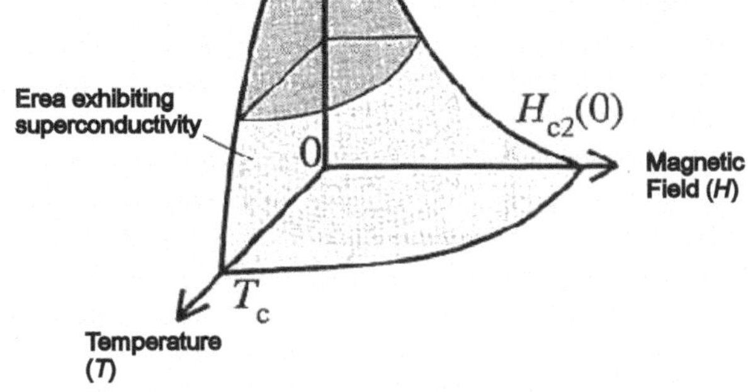

Fig. 8.15 Boundary condition of superconductor [9]

morphology and weak bonding strength between planes. Controlling the morphology increasing orientation is important to increase the critical current density.

When a superconductive material is placed in a magnetic field, the magnetic flux originally existed is ejected from the specimen. This is called the Meissner effect. In a high magnetic field, superconductive and normal conductive states coexist intruding quantum magnetic flux in the specimen. The superconductive state would disappear when the quantum magnetic flux moves on the Lorenz force. The effect to disturb the movement is called pinning effect and grain boundary, segregation particle, and dislocation act as pin.

8.6.3 Application of Superconductivity

Application of superconductivity lies in power transmission cable, superconducting coil, memory, and magnetic field sensor. For these applications, superconductive materials are processed to wire or thin film morphology. The requirement properties for power cable and coil are different, and then the figure of the wire would be changed to satisfy the requirement. For example, mechanical properties have to be improved so as not to expand the coil by the Lorenz force.

As a magnetic coil, magnetically levitated trains are famous application; however, the application field is expanding including magnetic separation and magnetic bearing. The Josephson tunnel junctions exhibit extremely high switching speed. The electrical noise level of superconductive material is extremely low which can be utilized to detect extremely low magnetic field. This device has already been practically used as SQUID.

References

1. H. Yanagida, M. Nagai (eds.), *Science of Advanced Inorganic Materials* (Shouko-do, Tokyo, 2000), p. 23
2. H. Yanagida (ed.), *Science of Ceramics*, 2nd edn. (Giho-do, Tokyo, 1981), p. 154
3. K. Uchino, *Ferroelectric Devices* (Tokyo, Morikita Syuppan, 2005), p. 171
4. N. Murayama, Ceramics **38**(11), 897 (2003)
5. S. Mizuta, K. Koumoto, *Ceramic Materials* (University of Tokyo Press, Tokyo, 1996), p. 92
6. A. Kishimoto, T. Kudo, Solid State Phys. **28**(5), 57 (1993)
7. A. Kishimoto, T. Kudo, Solid State Phys. **28**(5), 58 (1993)
8. J. Shimoyama, Ceramics **39**(3), 216–217 (2004)
9. J. Shimoyama, Ceramics **39**(4), 328 (2004)

Chapter 9
Optical Properties of Ceramics

Krisana Kobwittaya and Takanori Watari

Abstract This chapter provides an overview on essential principle of optical properties of ceramics which involves the phenomena in physics of the interaction between light and materials. Firstly, the light-matter interactions and how light behaves are described by considering the five key factors: scattering, absorption, transmission, reflection, and refraction. Also, this explanation includes three main groups of optical ceramics, transparent, translucent, and opaque, as material object and how to classify them. This chapter deals not only with physical backgrounds but also with giving the descriptions and examples of optical ceramics based on four crucial categories: transparent ceramic, single crystal, polycrystalline ceramic, and phosphor. Each sort is interpreted in detail with attribute, fabrication process, examples of material, and related applications.

Keywords Optical ceramics · Transparent ceramics · Single crystals · Polycrystalline ceramics · Phosphors

9.1 An Overview on Optical Properties of Ceramics

In general, the underlying phenomena of ceramic optical properties involve light-matter interactions which give rise to diverse behaviors of light, for example, light transmission, light reflection, and light absorption. Besides, these behaviors can be used to explain the nature of matter such as surface structure of materials.

The basic categorization of ceramics, which is directly related to optical properties, is usually described as transparent, translucent, and opaque. By this classification, glass is the most well-known material as a transparent ceramic. Alternatively, color in ceramics is one of the characteristics that occurs by light-matter interaction, and it can be changed by the number of point defects in ceramic structure or modified

K. Kobwittaya · T. Watari (✉)
Department of Science and Advanced Technology, Graduate School of Science and Engineering, Saga University, Saga, Japan
e-mail: watarit@cc.saga-u.ac.jp

J. Hojo (ed.), *Materials Chemistry of Ceramics*,
https://doi.org/10.1007/978-981-13-9935-0_9

by adding dopants or additives. For instance, glass is colorless material, but by adding the transition metals or rare earth elements, glass becomes colored material such as brown glass by doping Fe^{3+} and violet glass by doping Nd^{3+} [1, 2]. In recent years, the optical properties of ceramics have gained much attention because they are the essential knowledge for the development of various technologies, especially the advancement in light-emitting materials such as luminescent ceramics. With these three main issues, transmission, color, and light-emitting properties, the optical properties of ceramics lead to the considerable applications for employing in many technologies.

9.2 Basic Properties of Optical Ceramics

Theoretically, when light strikes a material, it may be scattered, absorbed, transmitted, reflected, or refracted (Fig. 9.1). These behaviors of light depend on the physical and chemical properties of that material. In case of any optical materials, there are five basic factors of optical phenomena: scattering, absorption, transmission, reflection, and refraction [3]. Also, these properties can be used to explain common peculiarities of optical ceramics, for example, transparency of ceramic and color in ceramic, including the optical coefficients such as refractive index.

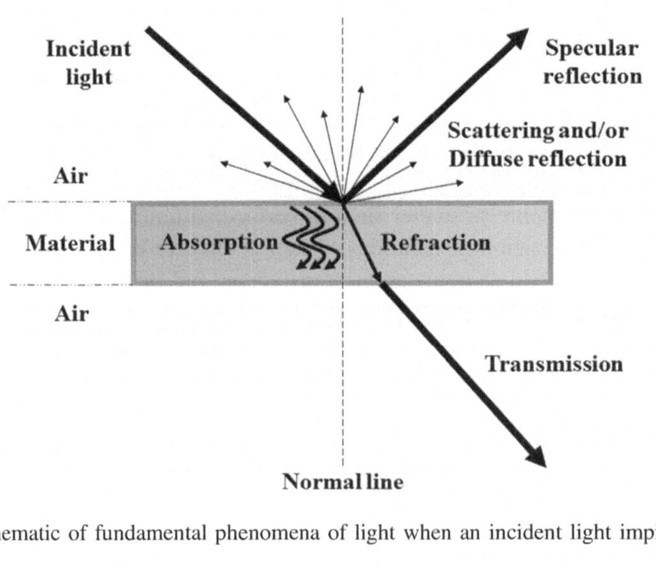

Fig. 9.1 Schematic of fundamental phenomena of light when an incident light impinges on the material

9.2.1 Scattering

The scattering is one of the initial phenomena, when a light wave meets any material object. Taking into account the structure of materials in microscopic level, they are different in particle sizes, surface roughness, or even pore sizes, which could affect the direction of light. Therefore, the scattering is a general physical process that occurs by the interaction of some part of light with localized non-uniformities in the medium, causing the light scattering in much different way from its original direction.

Normally, the question "why is the sky blue, albeit the sun is red" is a basis of thought for understanding fundamental scattering mechanism. This is due to a phenomenon called the Rayleigh scattering: light from the sun passing through the atmosphere is scattered by the individual molecules (preferably the particles that have a radius less than approximately a tenth of the wavelength of the light). Basically, the Rayleigh scattering refers to the scattering of light of air molecules, so this phenomenon is only involved in the particles that are much smaller than the wavelength of the light. Since light scattering is more probable for shorter wavelengths and the blue light has the shortest wavelength in visible region, the blue light from the sun is scattered more than other wavelengths, resulting in the blue color of the sky [4–7]. Within the visible range of light, the red light is scattered the least by atmospheric molecules. In general, at the time of sunrise and sunset, our eyes can detect red color in the sky because sunlight travels a long path through atmosphere to reach the eyes, and the blue light has been largely removed by long-distance travel, remaining mostly red light in the sky.

9.2.2 Absorption

The considerable point in this section is, after an incident light impinges on the surface of any materials, what happens to that light next? In previous section, light can be scattered, but sometimes light is absorbed and subsequently converted to the heat energy. This phenomenon is called absorption. The mechanism of light absorption is that the photons of the impinged incident light transfer its energy to atoms within the absorbing material. In most cases, the light absorption relies on the nature of atoms within the object. All atoms have specific vibrational frequency or also known as natural frequency. If an incident light interacts with those atoms with the same natural frequency, they will absorb the incident light energy, causing the heat of the object. In some cases, with the same mechanism mentioned above, the object exhibits the color due to the light absorption either, but this is due to the light absorption by electrons. Even though the absorbed light is responsible for the heating and coloring effects, these also depend on the amounts of light scattered, reflected or transmitted, including the original color of the material object. For example, when the colored optical materials are heated up by sunlight, dark materials lean to be hotter than light-colored material because they have low-light reflection, high absorption, and small or no light transmission, leading to more heat generation.

9.2.3 Transmission and Reflection

According to the relation of natural frequencies of light and atoms or electrons in material object, if the natural frequencies of both are compatible, the absorption of light in material takes place, but if not, the energy of light will reemit as a light wave again, resulting in the reduction of absorbed light or no absorption. For this reason, the reemitted light wave would be transmitted through the material or reflected by the atoms, depending on many factors, especially optical characteristics of material object. The simple way to explain the transmission and reflection of light is based on three basic groups of optical materials; transparent, translucent, and opaque.

The phenomenon which light goes through the object and then appears on the opposite site is called light transmission. On the other hand, some of light wave which goes back to the place where it came from is called light reflection. If light penetrates the object without much scattering or reflection, the object is a transparent material such as glasses and fused quartz. If the object allows the less light to go through it, which means that the scattering or reflection of light increases while the amount of transmitted light decreases, the object is a translucent material such as frosted glass and sunglasses. Hence, all things can be seen clearly through transparent materials, but not so well in the case of translucent materials. Nevertheless, there is another type of optical materials, which does not let any light go through. It is called opaque material such as traditional ceramics and glazed metals.

Reflection of light does not depend only on the type of material but also includes the surface roughness of material object. The reflection behavior of light would follow a predictable law known as the laws of reflection which are stated as follows:

1. The incident ray, the reflected ray, and the normal to the reflection surface, all lie in the same plane.
2. The angle of the incident ray is equal to the angle of the reflected ray when both angles are measured with respect to the mirror as a reflecting material.

Additionally, the appearance of color of any object can be interpreted by the transmission and reflection of light. When an object interacts with the light and subsequently transmits or reflects that light to human eyes, the color of that object will emerge. The light that makes human eyes detect the object color is in the visible region which consists of a wide range of light frequencies and wavelengths, leading to the various colors of object. So, each specific color of the object depends on the correspondence of natural frequency of electrons in object and each visible light frequency. Most colors in the environment are created by the objects that selectively reflect some wavelengths. This behavior is called selective reflection. For example, in the case of leaves, they absorb all visible light frequencies except for the frequency associated with green color, therefore, they will appear as green color.

Also, it is worth noting that the color depth of the objects is also due to the selective reflection of light. For instance, a green object appears bright-green color when irradiated with green light, but it looks dark-green color under the red light because it does not reflect red light well. In the case of the object that is transparent such as glasses and plastics, the color is created by selective transmission, meaning that the color of transparent objects is the color of light (only some wavelengths) that passes through the object. For example, a blue glass appears blue color because it allows the short wavelength light to transmit and appears blue.

9.2.4 *Refraction*

The objects that allow the light passing through are transparent and translucent materials. In general, a light wave has its specific direction, but when that light goes inside the object, its direction will be changed. This change of light direction inside the object is called refraction, which is the bending of a light wave direction. The physical reason of this phenomenon is about the dissimilar velocities of light in different mediums. Furthermore, the bending of light direction is also described by the laws of refraction which are almost the same as the laws of reflection.

1. The incident ray, the refracted ray, and the normal line, all lie in the same plane.
2. The direction of refracted ray is defined by the direction of the incident ray and the ratio of the velocities of light in the two materials:

$$\sin\theta_1/\sin\theta_2 = \upsilon_1/\upsilon_2 \quad (9.1)$$

where θ_1 is the angle of incident ray to the normal line, θ_2 is the angle of refracted ray to the normal line, and υ_1, υ_2 are the velocities of ray passing initial medium and second medium, respectively.

In principle, a light wave travels through a vacuum, passing at the speed of 3×10^8 m.s^{-1}. But when the light enters a medium, there are complex interactions between light and the atomic arrangement of the medium, leading to the change in the light velocity. The ratio of the velocity of light in a vacuum to that in the medium is known as the refractive index, *n*, which can be expressed by the simple equation:

$$n = \upsilon/\upsilon_m \quad (9.2)$$

where υ is velocity of light in a vacuum and υ_m is velocity of light in the material (medium). In addition, υ_m in any medium is always lower than υ; therefore, refractive index, *n*, is always greater than unity, except for in a vacuum ($n = 1$).

Factually, the velocity of light that changes from medium to medium has a significant effect on the wavelength of light. When a primary wave gets in the second medium, its wavelength changes. This change can be explained as follows.

Initially, each medium has specific refractive index that can be calculated by respecting to a vacuum as shown in Eqs. (9.3) and (9.4) for initial and second mediums, respectively.

$$n_1 = v/v_1 \tag{9.3}$$

$$n_2 = v/v_2 \tag{9.4}$$

Therefore, the relation of refractive index of two different mediums, in which both mediums are not vacuum, is

$$n_2/n_1 = v_1/v_2 \tag{9.5}$$

Subsequently, the velocity of any wave is calculated by

$$v = \lambda \mathrm{f} \tag{9.6}$$

where λ is the wavelength and f is the frequency of the wave. So,

$$n_2/n_1 = v_1/v_2 = \lambda_1 \mathrm{f}/\lambda_2 \mathrm{f} = \lambda_1/\lambda_2 \tag{9.7}$$

It is worth noting that the frequency is same because there are the same wave fronts travelling from first medium into second medium per unit time. Hence, n is a function of λ, and the variation of n with λ produces a spread of the light wave. The examples of relationship of n with λ are shown in Fig. 9.2.

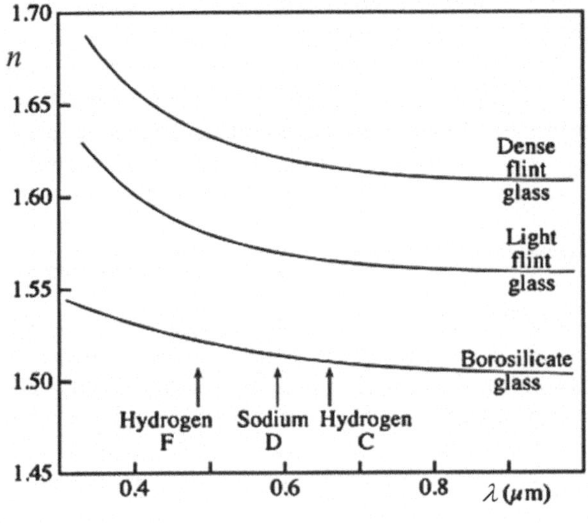

Fig. 9.2 Relationship between wavelengths (λ) and refraction index (n) for commercial glasses. Hydrogen F, Sodium D, and Hydrogen C refer to specific light sources, $\lambda = 486$, 589, and 656 nm, respectively. (Adapted with permission from Ref. [1]. Copyright (2013) Springer Science Business Media New York)

In addition, by the explanations mentioned above, the laws of refraction can be expressed in the terms of refractive index by the combination of Eqs. (9.1) and (9.5). The result shows the equation which is called Snell's law.

$$n_1 \sin\theta_1 = n_2 \sin\theta_2 \tag{9.8}$$

9.3 Transparent Ceramics

Nowadays, the most considerable optical determinant of any material is transparency (also known as pellucidity or diaphaneity). Transparency of materials is the key characteristic in current technologies, especially in the optical industries. This property can be obtained only in the materials in which the light can pass through them without being scattered [8, 9].

Some ceramics, both amorphous and crystalline, have existed as optically transparent materials in diverse forms from bulk solid-state components to high surface area forms such as thin films and coatings [10]. Typically, amorphous and crystalline ceramics can be briefly classified by the atomic arrangement. Amorphous (non-crystalline) ceramics, such as glasses which are the most well-known amorphous ceramic, are the materials in which the arrangement of atoms does not have long-range order. On the other hand, crystalline ceramics are the materials in which their atoms are intermittently arranged and located in a certain order in every part of the material. Also, the crystalline ceramics can be simply subdivided into single crystal and polycrystalline materials by considering atomic arrangement and the existence of grain boundaries [9, 11]. Therefore, this section provides underlying knowledges and examples of material for the three basic types of transparent ceramics: glass, single crystal, and polycrystalline ceramic.

9.3.1 Glasses

According to "The American Society for Testing and Materials (ASTM)" definition of glass, glass is an inorganic product of fusion that has cooled to a rigid condition without crystallizing [12]. Hence, glass is non-crystalline solid, or it would be called amorphous solid. In principle, glass materials can be mainly divided into three categories: metallic glass, polymeric glass, and ceramic glass [1]. Nevertheless, all glass materials in this section are relevant to only ceramic glass.

Glass components are the basis of all glass materials. In general, SiO_2 is the most well-known compound as a basic component of glasses, but there is not only SiO_2 but also the other components in order to form a usable glass. In the case of ceramic glasses, their components are mostly divided into two main important categories, network former and network modifier, which are classified based upon their roles and general physical appearance. Network formers such as SiO_2, B_2O_3, P_2O_5, and GeO_2 are the base structure of glasses in which their cations form coordination polyhedral in glass. To make glass useful for applications, the network modifiers such as Na_2O, K_2O, MgO, and CaO will be mixed together with network former in the initial stage of preparation. These network modifiers do not participate directly in the main glass network, but they are introduced into the glass to modify its properties. Each network modifier including its contents provides different modified properties of glass. Therefore, the selection of network modifier is significant for the preparation of a particularized glass for specific applications [1, 3]. In the following parts, the two examples of commonly used ceramic glasses, silicate (soda-lime) glass and borate (borosilicate) glass, will be described in detail.

9.3.1.1 Silicate (Soda-Lime) Glass

Silicate (soda-lime) glass is the simplest form of the glass, consisting of three main original compositions, SiO_2 (glass), Na_2O (soda), and CaO (lime), together with smaller amounts of MgO and Al_2O_3 in its crystalline system. The average refractive index of this glass is around 1.51–1.52. The soda-lime glass is the most widespread used glass in daily life because it is relatively low-price, chemically stable, practically hard, and enormously workable. With these properties, soda-lime glasses are also extensively applied for many applications, for example, windows in buildings, glassware such as bottles and jars, and lighting products such as bulb envelope, and they can be used economically to make plate, rod, and tube glasses as well. To date, one of the most interesting technologies of this glass is optical fiber for telecommunications due to the requirement of very low transmission loss over long distance in which this glass has lowest transmission loss compared to other optical glasses [1].

However, soda-lime glass is sometimes called soft glass due to its fragile nature which is the main disadvantage of this glass type. One of the serious drawbacks of soda-lime glasses is their relatively high thermal expansion because of the Na_2O composition. This leads to the dramatic increase in the volume of glass body, when this glass type is used at high temperature. Generally, glass which contains higher content of Na_2O shows very low thermal resistance, resulting in rapid change in temperature which is normally called a thermal shock phenomenon. Besides, this type of glass is much more difficult (or sometimes impossible) to repair, if it is broken into pieces [1, 3, 13, 14].

9.3.1.2 Borosilicate (Borate) Glass

Borosilicate glass (or sometimes called borate glass) is made by the combination of SiO_2, B_2O_3, and smaller amount of the other metal oxides such as Na_2O, K_2O, or Al_2O_3. The average refractive index of this glass is 1.47. The borate glass also shows low thermal expansion, high chemical stability, good insulation, and thermal shock resistance. The most familiar with this type of glass is in the form of laboratory apparatuses as heat-resisting ware with high resistance to strong acids such as beakers and flasks, which are better known under the trademark name, Pyrex™ by Corning or Duran™ by Schott Glass. In addition, borate glass can be found in various applications, for example, light bulbs, photochromic glasses, and high-intensity lighting devices. For the using as optical fiber, this glass is generally used for applications requiring light transmission over short distance such as in IR imaging devices and image bundles. The fascinating point of borate glass is the reparation. This glass is possible to repair, if it is broken because some borate glasses melt at very low temperatures (~500 °C), so they can be easily reformed or modified by joining together with other glasses [1].

9.3.2 Single Crystals

Transparent single crystal ceramics are a group of transparent crystalline solids, in which the atomic arrangement repeats periodically across its whole volume; besides, grain boundaries and defects are not present. Even though, inside single crystal structure, the atoms are well-organized throughout the material, the physical and mechanical properties may fluctuate, depending on the direction of the atomic arrangement. So, when the properties of a material change with different crystallographic alignments, the material is said to be anisotropic. For example, when a single crystal such as diamond is cleaved or cut, its atomic planes or crystal faces would be changed, resulting in the new direction of atomic alignments including the decrease in atomic bond strength and density of bonds [11].

Transparent single crystal ceramics have been used in highly specific applications such as high-quality optical products, laser technologies, and piezoelectric sensors because these groups of applications require the materials which do not show any absorption in a large-range between infrared to ultraviolet and are appropriate to work under extreme conditions. Albeit single crystals show very high transparency as well as very good chemical and thermal stability, these materials that can be produced are limited because of two huge problems. First, the single crystals are typically grown from the melt or by frame processes, and the machining requires sophisticated instruments, causing to the very expensive task. Second, the single crystals have shapes controlled more by their lattice structures and less by processing conditions which means that the products are restricted to simple shape manufacture. Hence, the production of single crystals to meet the requirement of any application is a difficult task which largely limits their increase in production and, therefore, their wide range of applications [9, 15]. The two examples of the most familiar transparent single crystal ceramics include synthetic sapphire (Al_2O_3) and single crystal yttrium garnet ($Y_3Al_5O_{12}$, YAG), which will be discussed in the following parts.

9.3.2.1 Synthetic Sapphire (Al_2O_3)

Synthetic sapphire is a single crystal of corundum aluminum oxide (Al_2O_3), also commonly known as alumina, alpha alumina (α-Al_2O_3), and single crystal aluminum oxide. Corundum Al_2O_3 has a hexagonal crystal lattice with cell parameter $a = 4.754$ Å and $c = 12.99$ Å. The O^{2-} ions are organized in close-pack hexagonal alignment, with the Al^{3+} ions occupying two-thirds of the octahedral interstitial positions [16]. In addition, it has a density of 3.95 g/cm^3, with melting point of 2072 °C and boiling point of 2977 °C [9]. The most preferable manufacturing method for single crystal Al_2O_3 is the EFG (edge-defined film-fed growth) method because it is possible to grow inexpensive materials (e.g., raw Al_2O_3 powders) with high crystal quality [11].

Hypothetically, sapphire is Al_2O_3 in the uncontaminated form without porosity or grain boundaries, making it apparently dense. Besides, it is a very strong material because Mohs hardness scale is 9 out of 10 (the material which shows maximum Mohs hardness scale is diamond). As a result, sapphire is an adaptable material with a lot of applications due to its excellent properties such as outstanding optical properties, superior mechanical properties, physical and chemical stability, good thermal conductivity, and high heat resistance, which make sapphire as a preferable choice of transparent materials for being used in high performance applications. Albeit sapphire shows outstanding properties, the machining of this material is quite difficult, inordinately expensive, and time-consuming. For the applications, sapphire is widely used, for instance, as substrate materials for light-emitting diodes (LEDs) and thin films, as semiconductor process equipment, and as optical products for LCD projector with high picture quality [11, 17]. In the case of colored sapphires (typically found in general form of natural sapphire), the single crystal Al_2O_3 is pure crystal which is transparent and colorless, but it can become blue (blue sapphire), if the single crystal Al_2O_3 contains impurities such as Ti and Fe and red if the impurity is Cr. The red sapphire is one of the well-known gemstones, named as ruby [11]. When the single crystal Al_2O_3 demonstrates the color, they are not single crystalline structure but become polycrystalline structure due to the existence of impurities, resulting in the change in crystalline alignment.

9.3.2.2 Single Crystal Yttrium Aluminum Garnet ($Y_3Al_5O_{12}$, YAG)

In the Y_2O_3-Al_2O_3 composite system, three phases are known, monoclinic phase ($Y_4Al_2O_9$, YAM), perovskite phase ($YAlO_3$, YAP), and cubic phase ($Y_3Al_5O_{12}$, YAG), and the formation and preparation temperature of each phase is shown in the following equations [18]:

$$2Y_2O_3 + Al_2O_3 \rightarrow \text{YAM} \ (900 - 1100^\circ\text{C}) \tag{9.9}$$

$$YAM + Al_2O_3 \rightarrow 4YAP\ \left(1100 - 1250^{\circ}C\right) \tag{9.10}$$

$$3YAP + Al_2O_3 \rightarrow YAG\ \left(1400 - 1600^{\circ}C\right) \tag{9.11}$$

In the case of cubic phase, it is named as garnet because there are three cationic sites: tetrahedral, octahedral, and dodecahedral. The YAG materials have been broadly studied over many decades because of their mechanical strength, chemical stability, resistance at very high temperature, and unique homogenous optical properties such as transparent from below 300 nm to beyond 4 microns [9, 19]. Typically, simple formed YAG is one of the materials that can be used for making synthetic gemstones, or it is called YAG jewelry.

With YAGs' excellent properties, single crystal YAG is interested and performed by Czochralski process, with crystal growth conditions at pull rates of 0.4 mm/h and temperatures of nearly 1900 °C. By this process, the product demonstrates high performance over traditional transparent materials for various applications, for example, single crystal YAG fibers, in which their physical properties are greater to those of amorphous silica with much higher output powers [20, 21]. Further, single crystal YAGs have been developed and then can dominate the commercial transparent materials for solid-state laser applications because single crystal YAGs are able to accept trivalent laser activator ions of either rare earth or transition metals and can be grown with very low strain. The most famous and well-known application by using single crystal YAGs doped with rare earth metals is Nd:YAG laser. However, transparent single crystal YAG not only is a significant material for laser applications but also plays an important role as an excellent host material for light-emitting materials [22, 23].

9.3.3 Polycrystalline Ceramics

Many transparent ceramics are in the form of single crystal because their properties are suitable for use as transparent material more than any material, but the manufacturing process of single crystal materials to meet the requirements of any application includes difficult and expensive tasks which mostly limit their increase in production and, therefore, their wide range of applications. Formally, polycrystalline materials are the promising candidate for use as transparent material because of their properties such as high mechanical strength, physical and chemical stability, and good thermal resistance which are similar to single crystals. Besides, they can be fabricated in versatile shapes, large sizes, and more importantly cost-effectiveness.

In theory, polycrystalline ceramics are a group of crystalline solids, in which they contain many grains where the orientation of those grains is usually different from each other. In comparison with single crystals, polycrystalline materials have more complex microstructure which consists of, for example, grain boundaries, residual pores, secondary phase, double refraction granule, rough surface, and many point defects. Most of these structural characteristics are scattering center which directly affects optical properties, especially transparency of the materials. To make respectable transparent polycrystalline materials, these should be poreless and have relatively clear grain boundaries and crystals to minimize light scattering. Up to now, the best way to make transparent polycrystalline ceramics is sintering the material in high-temperature furnace. The sintering mechanism involves diffusional transport of matter along specific paths which simplifies the structure of polycrystalline ceramics. Unfortunately, the sintering method cannot remove all pores to make perfect transparent polycrystalline ceramics. Nowadays, the removal of remaining pores represents a most important challenge and has been the issue of several studies [24, 25]. Taking into account the transparent material applications, transparent polycrystalline ceramics have become a priority material for various technologies such as lasers, optical lenses, lamp envelopes, and transparent armors, because of their advantages over single crystals and glasses [8]. The two examples of the most familiar transparent polycrystalline ceramics include polycrystalline alumina (Al_2O_3) and polycrystalline zinc sulfide (ZnS), which will be discussed in the following parts.

9.3.3.1 Polycrystalline Alumina (Al_2O_3)

Polycrystalline alumina is one of the two crystalline Al_2O_3 (the other one is single crystal Al_2O_3), also known as Al_2O_3 ceramic. Typically, traditional Al_2O_3 ceramics are produced by sintering Al_2O_3 powder, and the obtained products are opaque due to the light scattering by many pores and grain boundaries. However, this material can be transparent, if it is prepared by suitable condition and technique. To make Al_2O_3 ceramic as transparent material, the considerable point is grain size and residual porosity reduction. Hence, the optical properties of Al_2O_3 ceramics can be improved by increasing their purity and density and by controlling their microstructures. For this reason, various methods have been employed to control the grain size and minimize the residual porosity.

After the first transparent polycrystalline Al_2O_3 was proposed and patented by Coble [26], many researchers reported various sintering techniques to produce transparent polycrystalline Al_2O_3 ceramics such as microwave sintering [27], pulse electric current sintering (PECS) [28, 29], and spark plasma sintering (SPS) [30, 31]. The fine-grained transparent polycrystalline Al_2O_3 ceramic is one of the products that have attracted much attention because it shows a significant improvement in mechanical strength and optical transparency. This material is prepared by two-step sintering, hot pressing (HP) followed by hot isostatic pressing (HIP), at firing temperature 1150 to 1400 °C. It is reported that this material has strength of up to 400–600 MPa and high in-line transmission of up to 60% for visible light. Additionally, doping is found as one of the effective techniques for the development of transparency of polycrystalline Al_2O_3 ceramics. For example, after doping and sintering, MgO doped Al_2O_3 ceramic has nearly full density and in-line transmission of 40–50% for visible light. This improved transmission is caused by the segregation of MgO into Al_2O_3 grain boundaries, resulting in the reduction of porosity and grain size and the homogenization of the microstructure.

Basically, Al_2O_3 ceramics demonstrate many interesting properties, such as high hardness and strength, and excellent corrosive resistance. With these original properties and the enhanced optical and mechanical properties by sintering techniques, the transparent polycrystalline Al_2O_3 ceramics with submicron grain size are one of the hardest materials among all transparent materials, even including single crystal Al_2O_3. Therefore, transparent polycrystalline Al_2O_3 ceramics including the modified one are believed to be a promising alternative for the replacement of single crystal Al_2O_3, and they are a hopeful candidate for various applications, for instance, electromagnetic windows, high-temperature light-transmitting materials, and envelopes of high-pressure (HP) metal-halide lamps [8, 9].

9.3.3.2 Polycrystalline Zinc Sulfide (ZnS)

Zinc sulfide (ZnS) exists in two common polymorphs: cubic (zinc blende, also known as sphalerite) or hexagonal (wurtzite). The lattice structures of both types have these characteristics; stoichiometry of Zn to S is 1:1, and each ion has a coordination number 4 in tetrahedral configuration, but the difference between both types is that zinc blende is based on a face-centered cubic (FCC) lattice of anions, whereas wurtzite is derived from a hexagonal close-packed (HCP) array of anions. Typically, zinc blende is the main form of ZnS found in nature, and it is usually black because of the impurities, especially irons, but the pure form is white, and it is usually used as pigment. However, ZnS not only is found in nature but also can be synthesized by various methods: evaporation, sublimation, high-pressure growth from molten ZnS, sintering and hot pressing of polycrystalline ZnS, and chemical vapor deposition (CVD). In the case of polycrystalline ZnS fabrication as transparent material, CVD method is extensively used because the products are accurately controlled in either stoichiometry of compositions or purity of material. Furthermore, this technique is relatively attractive when the material requirement is high physical and chemical perfection with near-net shaping. The reaction for growing ZnS by CVD method is shown in Eq. (9.12).

$$Zn_{(g)} + H_2S_{(g)} \rightarrow ZnS_{(s)} + H_{2(g)} \tag{9.12}$$

The CVD ZnS shows very low transmission in the visible range due to the appearance of Zn-H absorption band. Hence, the posttreatment is required to improve the transmission properties as well as the elimination of Zn-H absorption band. The hot isostatic pressing (HIP) is widely used as the posttreatment for CVD method because HIP can improve the grain alignment and grain size and reduce remaining porosity. Nowadays, the combination of CVD and HIP methods is the underlying process in many manufactures to produce the transparent polycrystalline

ZnS for a variety of applications such as infrared windows, lenses, and domes [32, 33].

9.4 Luminescence

In principle, the luminescence is described as a phenomenon that the electronic states of substance are excited by external energy source and further that substance releases the excited energy in the form of light at various wavelengths [34]. The luminescence phenomena can be grouped into various categories, depending on the type of excitation source. For example, cathodoluminescence (CL) is excited by cathode rays or electron beams; thermoluminescence (TL) by thermal activation after initial irradiation such as α, β, γ, UV, or X-rays; photoluminescence (PL) by electromagnetic radiation after photon absorption; and electroluminescence (EL) by electric influences [35, 36]. Additionally, luminescence can be divided into fluorescence and phosphorescence by the basis of duration of emission or decay time. The fluorescence denotes as an emission of short decay time less than 10^{-8} s, and its emission is seen to be happening concurrently with the absorption of radiation, and it will be stopped instantaneously, if the excitation source is removed. On the other hand, the phosphorescence signifies as an emission of long decay time more than 10^{-8} s, and its emission is seen to go on for some time even if excitation source has been removed. However, it is clear that the fluorescence is essentially temperature independent, whereas the phosphorescence shows strong temperature dependency [37, 38].

The basic concept of luminescence mechanism is shown in Fig. 9.3. This figure shows two types of energy transition after the energy is absorbed and subsequently excited to the excited state: one is the radiative transition that would be called luminescence, and the other one is non-radiative transition that occurs due to the

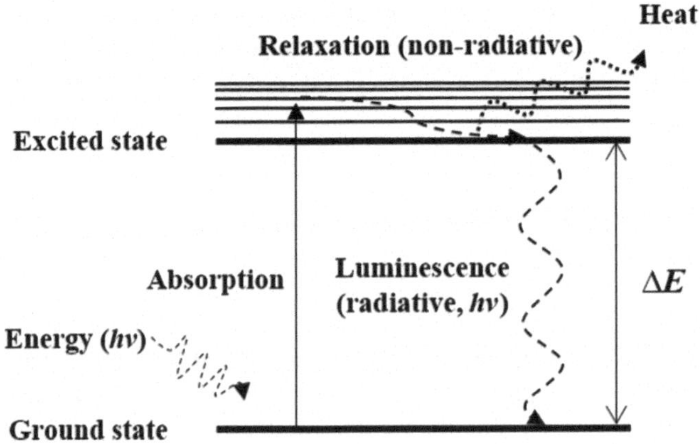

Fig. 9.3 Simplified basic luminescence mechanism based on excitation/emission processes

lattice vibration, resulting in the energy transfer in the form of heat. Nevertheless, the non-radiative does not have the important role in luminescence yet. In general, the productive luminescent material is directly related to radiative transitions which should dominate over the non-radiative transitions.

Basically, luminescent materials emit a variety of light colors in the visible spectrum. The color of light depends on the energy gap (ΔE) between excited state and ground state. In principle, the energy gap can be calculated by using Planck's Eq. (9.13)

$$E = hv = hc/\lambda \tag{9.13}$$

where E is the energy of photon, h is Planck's constant (6.63×10^{-34} J.s), c is velocity of light (3×10^{8} m.s^{-1}), λ is wavelength (m), and v is frequency (Hz). Normally, the human eyes can detect the visible light in the visible range (400–700 nm). Therefore, if ΔE is around 2.95 eV, which corresponds to wavelength 400 nm, the human eyes can detect violet color.

Luminescent material, also called a phosphor, is a solid that changes certain types of energy into electromagnetic radiation over and above thermal radiation. Most luminescent materials take the advantages of a wide optical gap in their electronic band structure or otherwise, the presence of defects or impurities which plays a significant role as the localized electronic states. The luminescence emission exhibited by a luminescent material is regularly in the visible range but could also occur in ultraviolet (UV) or infrared (IR) regions. In the field of luminescent ceramics, there are three underlying types of the material which comprise of phosphor, upconversion phosphor, and long persistent phosphor. Therefore, this section will provide the general description of phosphors with the emphasis on luminescence mechanisms and their characteristics. Besides, the aspects of the three fundamental types of phosphor are briefly defined in the following sections.

9.4.1 Phosphors

Phosphor is one of the luminescent materials that can absorb energy from the incident radiation and then emits the photons in the form of light after energy transfer processes. Its emission usually falls into the visible range, but in some phosphors, they can exhibit invisible luminescence such as ultraviolet or infrared radiation. In principle, the phosphors are classified into organic and inorganic phosphors. Organic phosphors involve specific molecules, while inorganic phosphors involve certain lattice structures. Most luminescent ceramics are inorganic phosphor which consists of an inert imperfect host crystal lattice to which some impurity ions called dopants are intentionally added. Typical dopants are transition or rare earth metals which are the most common optically active impurities. There are two kinds of dopant in phosphor materials. One is employed as the luminescent center which emits the radiation and is called an activator. The other one harvests the excited energy efficiently and then transfers to the activator and is called a sensitizer which would be added to the phosphors, if the activator cannot be excited, for example, because of the parity forbidden transitions [38].

9.4.1.1 Fundamental Emission and Excitation Mechanisms of Phosphor

Fundamentally, the luminescence of phosphors takes place by the absorption of energy at the activator site, followed by energy relaxation, and then they return to the ground state by emission of photon with energy *E*. In addition, the energy absorption of phosphors can occur via the host lattice or by intentionally doped impurity ions.

According to Fig. 9.4, the schematic shows how phosphor emits the light. There are two kinds of process based on energy absorption performance of activator ion. As shown in Fig. 9.4(a), the gray circle (A) is signified as an activator ion which is enclosed by the host lattice (white circle). The luminescence process is a release of energy in the form of emission (radiative) and some energy in the form of heat (non-radiative) due to the internal oscillations of material that normally occur through lattice structure. In the host material that contains only activator ion, the external energy directly excites this ion; subsequently, its electrons absorb energy and are provoked into an excited state. After that, the energy from that excited state relaxes back to the ground state by the emission of photons. In the Fig. 9.4(b), when the activator ion (gray circle (A)) can absorb small energy or it means weak energy absorption, the second ion will be added in which this ion is generally known as the sensitizer ion (gray circle (B)). The sensitizer ion can absorb more energy and then transfers its energy to the activator ion for emitting the emission. In addition, the sensitizer ion can exhibit the emission by itself, but in most of all cases, the strong emission usually occurs via activator ion because of the energy transfer from sensitizer ion to activator ion.

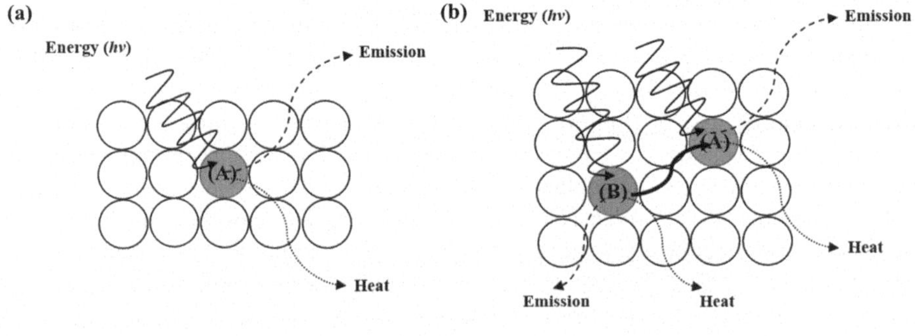

Fig. 9.4 Schematic diagram of how phosphor emits the light with (**a**) direct excitation from the activator (A) and (**b**) indirect excitation from the activator (A) and sensitizer (B)

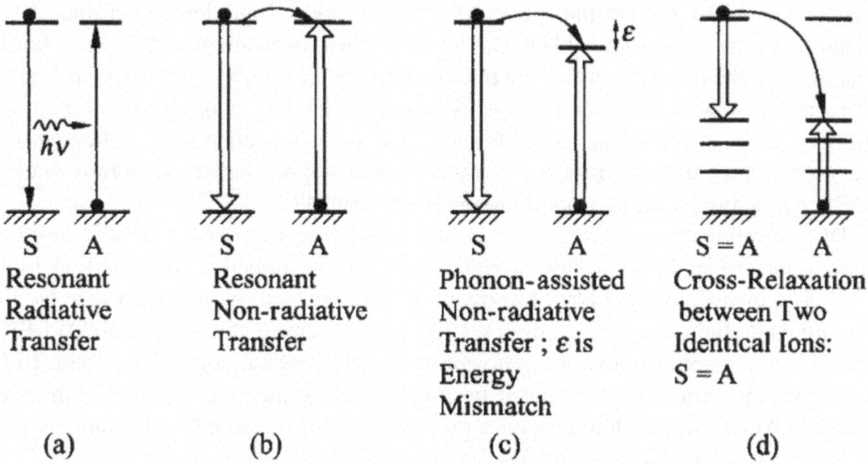

Fig. 9.5 Schematic diagram of phosphor with different ET processes between two ions, (**a**) resonant radiative ET, (**b**) resonant non-radiative ET, (**c**) phonon-assisted ET, and (**d**) cross-relaxation. (Reproduced with permission from Ref. [39]. Copyright (2005) Tsinghua University Press and Springer-Verlag Berlin Heidelberg)

9.4.1.2 Energy Transfer Processes in Phosphor

Energy transfer (ET) is a process in which the sensitizer and activator ions show physical interaction. As shown in Fig. 9.5, there are four basic mechanisms involved in ET processes between sensitizer and activator ions: (1) resonant radiative ET, (2) resonant non-radiative ET, (3) phonon-assisted ET, and (4) cross-relaxation. In general, an efficient ET process requires a vital spectral overlapping between the emission spectrum of sensitizer and the absorption spectrum of activator. This is a very significant point of ET process which takes place in a variety of luminescence mechanisms of phosphors. For resonant radiative ET (Fig. 9.5(a)), the photons emitted by the sensitizer ions are absorbed by any activator ion within a photon travel distance, and the sensitized emission lifetime does not change with the activator concentration. In contrast to the resonant radiative ET, the resonant non-radiative ET (Fig. 9.5(b)) occurs without emission of photons and is often accompanied by an important decrease in the sensitized emission lifetime against activator concentration [39, 40].

By concerning the energy level mismatch (ε) between sensitizer and activator ions, some ET occurs with emission or absorption of phonons to compensate the energy mismatch. This mechanism is called phonon-assisted ET (Fig. 9.5(c)). Moreover, the ET also involves the distance from sensitizer to sensitizer, activator to activator, or sensitizer to activator. If the distance of two ions is so close and the excited energy of one ion is in higher excited state, the photons which return to the lower energy level can transfer its energy to the neighboring ions. This ET is called cross-relaxation and usually refers to downconversion (DC) ET (Fig. 9.5(d)). With this DC ET model, the DC phosphors have been proposed and developed since 1957 as the first type of phosphor [41]. The luminescence mechanism of DC phosphor is described as the process, where one photon with a high energy is cut to obtain one or more photons with a lower energy. Nowadays, the DC phosphor is sometimes shortly called as phosphor. In addition, when the obtained photons after cutting are more than one, this process is referred and approached to quantum cutting (QC) or quantum splitting (QS) in many literatures [42].

DC phosphors have been developed and investigated for two decades for their use in the solar cell applications, for example, Pr^{3+}/Yb^{3+} co-doped $Y_2Si_2O_7$ [43] or Er^{3+}/Yb^{3+} co-doped $CaMoO_4$ [44]. Moreover, DC process is the essential knowledge for making white light-emitting diodes (WLEDs). For instance, the commercial WLEDs usually are phosphor-converted conventional LEDs, which consist of a blue LED (e.g., gallium nitride (GaN)) coated by a yellow phosphor (e.g., yttrium aluminum garnet (YAG)). General luminescence processes of this material for generating white emission are clarified as follows. When a blue LED which is excited by short wavelength excitation source emits blue emission, a yellow phosphor will absorb that blue emission and then down converts to yellow emission. So, the combination of blue emission from blue LED and yellow emission from yellow phosphor produces white emission.

9.4.2 *Upconversion Phosphors*

Upconversion (UC) phosphor is a solid luminescent material in which the process involves the absorption of two or more photons of low energy with subsequent emission of higher energy photons. In the easy way to explain, UC process is a way to convert long-wavelength radiation into shorter wavelength radiation. So, the UC process is remarked as the reversal process of the downconversion.

9.4.2.1 Luminescence Mechanisms of Upconversion Phosphor

UC process is mainly divided into three classes: excited state absorption (ESA), energy transfer upconversion (ETU), and photon avalanche (PA). These processes involve the population of a highly excited state by the sequential absorption of two or more photons by metastable, long-lived energy states (Fig. 9.6). The first UC process is the two-step absorption in which a single ion absorbs two subsequent photons (Fig. 9.6(a)). Firstly, the single-emitting ion is excited by the excitation source from the ground state (G) to an intermediate energy level (E1); this step is called ground state absorption (GSA). Secondly, the absorption occurs from E1 to higher state E2, this transition is called excited state absorption (ESA), but this second absorption must have high enough excitation energy within the lifetime of the energy in E1 to prevent its decay to the ground state. Thereafter, UC

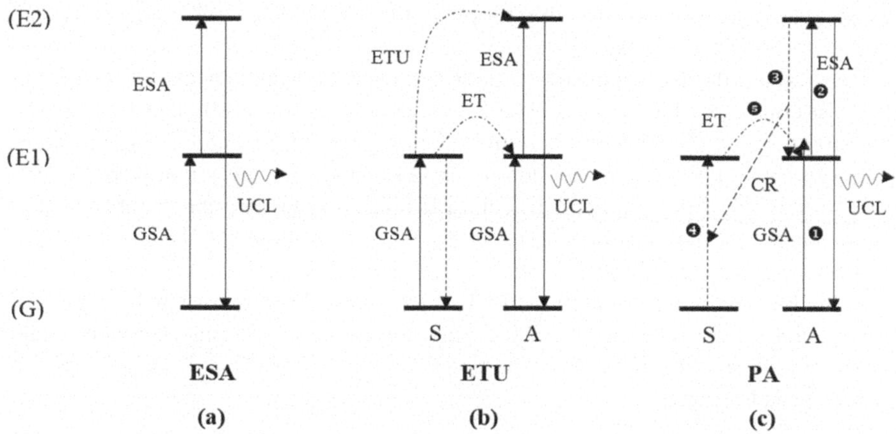

Fig. 9.6 Principle energy diagrams of various upconversion processes for generating upconversion luminescence (UCL)

luminescence (UCL) takes place through radiative transition from E2 to G. In general, this mechanism can occur in the single ion-doped materials with low doping level.

Figure 9.6(b) shows the energy transfer upconversion (ETU). This mechanism is quite different from ESA because ESA takes place within a single ion, while ETU includes two neighboring ions (sensitizer (S) and activator (A) ions). In the ETU, each ion absorbs a pump photon of the same excitation energy by GSA, populating energy to metastable level E1. Subsequently, the energy at E1 of S transfers to A by two ways, ET and ETU, while the excess energy at E1 of S relaxes back to the ground state G. Further, the energy at E1 of A is promoted to higher excited state (E2) by ESA. Afterward, UCL takes place through radiative transition from E2 to G as well as the ESA model. Since this system consists of two neighboring ions, the most significant point is average distance between them which has a high effect to the UC efficiency of ETU process. This distance can be adjusted by varying dopant concentrations, and the suitable concentration is needed to acquire the proper distance and high UC efficiency.

Figure 9.6(c) shows the photon avalanche (PA). This process features an unusual pump mechanism in which the system requires high pump power above the threshold value. The process begins with the population of photon from G of A to E1 of A by non-resonant weak GSA (❶), followed by a resonant ESA to populate the higher excited E2 level (❷). Subsequently, the photon at E2 relaxes back to E1 (❸), during this transition, a cross-relaxation energy transfer (CR) (also called ion pair-relaxation) occurs from the E2 → E1 transition of A to ground state G of S (or expressed as ❸ to ❹), and then the photon is populated from G of S to E1 of S. After that, that photon occupies the intermediate energy level E1 of S, resulting in the ET from E1 of S to E1 of A (❺). Further, E1 of A is populated again to E2 by ESA to continuously initiate cross-relaxation, producing the looping process, or it is called avalanche process. By this looping, the UCL increases dramatically and produces strong UC emission.

Considering the UC efficiency of these three luminescence mechanisms, ESA is the least competent UC process. The PA process is the most efficient UC, but the emission response is delayed due to the CR processes, numerous looping processes with a rise time up to a few seconds, weak GSA, and the high pump power over a certain threshold limit required. In contrast, ETU is constant, and pump power is independent and thus has been extensively used to propose highly efficient UCL over the past decade [45, 46].

In the theoretical understanding of UC mechanism, this process can be explained by considering the relationship of UC emission intensity (I) and excitation pump power (P). The intensity I is proportional to the n power of P, which can be expressed as follows:

$$I \propto P^n \tag{9.14}$$

where n is the number of pump photons per one photon emitted. To obtain the n value, a plotting of log I versus log P yields a straight line with slope n [45]. Principally, under steady-state excitation, the n value is in the specific range value and less than the maximum theoretical n value for each possible UC process, two-photon process ($1 < n \leq 2$) or three-photon process ($2 < n \leq 3$), because of the saturation effect in the UC luminescence intensity caused by the competition between a few processes, for instance, the UC process and non-radiative energy transfer between dopant ions for the depletion of the intermediate excited states [47, 48].

9.4.2.2 Structure of Upconversion Phosphor

The UC phosphors generally consist of an inorganic crystalline host and dopant ions. In most cases, the dopant ions are rare earth ions which are commonly in the form of localized luminescent centers. Due to the discrete energy states of these ions, rare earth doped upconverters exhibit superior infrared to visible UC performance. Hence, UC luminescence can be expected in principle from most rare earth doped host crystal materials, but the suitable UC luminescence only takes place by using small number of well-designated dopant-host combinations.

Ideal host material should have a low lattice phonon energy, which is a requirement to decrease non-radiative loss and increase the radiative emission. Basically, it seems that heavy halides group (chlorides, bromides, and iodides) are the good UC host materials since they exhibit very low phonon energy (less than 37 meV), but they have many disadvantages such as poor chemical resistance, thermal unsteadiness, and high cost. Therefore, this group is difficult to use in the practical applications. Then and now, oxide is one of the most investigated UC phosphor host materials due to their good properties such as chemical stability and phonon energy around 68 meV. However, oxide may not be an appropriate host material for UC phosphor because its phonon energy is relatively high. To date, fluoride is the most promising UC phosphor host materials because they show chemical stability, low phonon energy (~43 meV), and high optical transparency over a wide wavelength range. Among many fluoride host materials, the crystalline $NaYF_4$ is one of the most efficient host lattices for UC phosphor. Also, it is widely used as UC host because of its two modification forms, a high-temperature cubic phase (α-$NaYF_4$) and a low-temperature hexagonal phase (β-$NaYF_4$), which can be made a variety of UC phosphors, for example, transparent UC phosphor or UC phosphor in nanoscale.

The multiple metastable levels requirement for UC processes makes rare earth ions well-suited for making efficient UC phosphor. To generate high performance UC luminescence, the energy difference between each excited level and its lower-lying intermediate level (ground level) should be close to the excitation energy to facilitate the photon absorption and energy transfer steps involved in the UC processes. According to two kinds of rare earth ions doped in the UC phosphors, sensitizer and activator, trivalent ytterbium ion (Yb^{3+}) is the most examined as a sensitizer because it possesses a simple energy level scheme with only one excited 4*f* level of $^2F_{5/2}$. Besides, the absorption band of Yb^{3+}, which locates around 980 nm due to the $^2F_{7/2} \rightarrow {}^2F_{5/2}$ transition, has a larger absorption cross section than that of other rare earth ions. Additionally, the $^2F_{7/2} \rightarrow {}^2F_{5/2}$ transition of Yb^{3+} ion is well resonant with many *f*-*f* transitions of typical rare earth activator ions (Er^{3+}, Ho^{3+}, and Tm^{3+}), leading to the facilitating resourceful ET from Yb^{3+} to these ions. These characteristics make Yb^{3+} ion appropriate for use as an UC sensitizer. The content of sensitizer is normally kept high (around 20 mol%) in co- or tri-doped ions system, while the activator content is relatively low (less than 2 mol%) to minimize energy loss from cross-relaxation. In this section, the underlying UC mechanism of well-known couples of rare earth ions, Yb^{3+}-Er^{3+}, Yb^{3+}-Tm^{3+}, and Yb^{3+}-Ho^{3+} (Fig. 9.7), will be discussed as follows.

In the Yb^{3+} sensitized Er^{3+} UC luminescence (Fig. 9.7(a)), the Yb^{3+} is excited from $^2F_{7/2} \rightarrow {}^2F_{5/2}$ energy level by excitation source around 980 nm, and at the same time, some energy may relax back to the ground state $^2F_{7/2}$ level as radiative or non-radiative transitions. Considering the long lifetime of the excited $^2F_{5/2}$ level (typically one millisecond), Yb^{3+} may well transfer the excitation energy to an Er^{3+} with higher probability than decaying the excited energy to its ground state [50]. Subsequently, Er^{3+} is populated to $^4I_{11/2}$, $^4F_{9/2}$, $^4S_{3/2}$, and $^4F_{7/2}$ levels due to the absorption process and energy transfer from Yb^{3+}. After that, the energy in each energy level of Er^{3+} may relax back to $^2H_{11/2}$, $^4S_{3/2}$, or $^4F_{9/2}$ levels in the form of non-radiative transition. The UC emission is customarily assigned to the following transitions: green emission centered around 520 and 540 nm wavelengths by $^2H_{11/2} \rightarrow {}^4I_{15/2}$ and $^4S_{3/2} \rightarrow {}^4I_{15/2}$ transitions of Er^{3+}, respectively, and red emission centered around 660 nm wavelength by $^4F_{9/2} \rightarrow {}^4I_{15/2}$ transition of Er^{3+}.

In the Yb^{3+} sensitized Tm^{3+} UC luminescence (Fig. 9.7(b)), under the 980-nm excitation, Yb^{3+} is excited from $^2F_{7/2} \rightarrow {}^2F_{5/2}$ energy level. Further, the energy in $^2F_{5/2}$ energy level of Yb^{3+} is transferred to the Tm^{3+}, and then Tm^{3+} is populated to the corresponding excited energy level due to the absorption process. The emission bands centered at around 450, 475, 645, 690, and 800 nm wavelengths are attributed

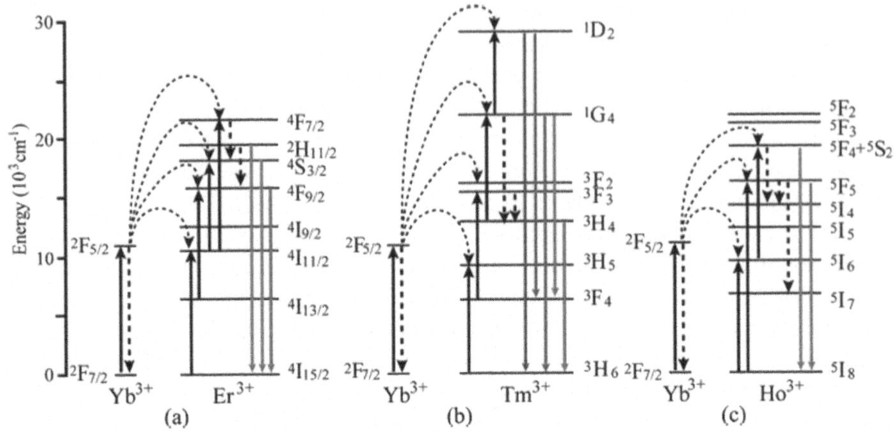

Fig. 9.7 Upconversion mechanism of the material, containing (**a**) Yb^{3+}-Er^{3+}, (**b**) Yb^{3+}-Tm^{3+}, and (**c**) Yb^{3+}-Ho^{3+}. (Adapted with permission from Ref. [49], Copyright (2015) American Chemical Society)

to $^1D_2 \rightarrow {}^3H_6$ (ultraviolet), $^1D_2 \rightarrow {}^3F_4$ (violet), $^1G_4 \rightarrow {}^3H_6$ (blue), $^1G_4 \rightarrow {}^3F_4$ (red), and $^3H_4 \rightarrow {}^3H_6$ (NIR) transitions of Tm^{3+} ion, respectively.

In the Yb^{3+} sensitized Ho^{3+} UC luminescence (Fig. 9.7(c)), the Yb^{3+}/Ho^{3+} co-doped materials exhibit two main emissions, green and red luminescence, due to the following transitions: green emission centered at around 550 nm wavelength due to the $^5F_4 + {}^5S_2 \rightarrow {}^5I_8$ transition and red emission centered at around 670 nm wavelength due to $^5F_5 \rightarrow {}^5I_8$ transition [45, 46, 49, 51].

9.4.2.3 Applications

Over the past decade, the researches on UC phosphors have made gigantic progresses in many applications such as solar cells, displays, and photocatalysis. However, the most interesting applications of UC phosphors are in the field of biomedical such as imaging, sensing, and cancer therapy, because of their outstanding merits, for example, multi-colored emission capability under single wavelength excitation, high signal-to-noise ratio, low cytotoxicity, and high chemical and photostability [49]. In this part, the most popular types of biomedical technology, in vivo detection and in vivo imaging, including some examples of the UC phosphor used in these applications will be discussed as follows.

In the in vivo detection applications, UC phosphors have been used as a luminescent reporter in a variety of in vivo assays, including immunoassays, bioaffinity assays, and DNA hybridization assays, which offers dramatically enhanced signal-to-noise ratio and thus improved detection limits compared to conventional reporters [46]. For example, Hampl et al. [52] reported a detection limit of 10 picogram human chlorionic gonadotropin in a 100-μL sample by using submicron-sized Y_2O_2S:Yb/Er particles in immunochromatographic assays. Simultaneous detection by multiple analysis using multi-colored UC nanocrystals was demonstrated by Niedbala et al. [53]. In a lateral flow-based competitive inhibition assay, drugs containing low concentrations of amphetamine, methamphetamine, phencyclidine (PCP), and opiates were detected in a single multiplexed assay strip. To achieve the multiplexed detection, antibodies for PCP and amphetamine were conjugated to green-emitting particles, while methamphetamine and morphine were conjugated to blue-emitting particles.

For the in vivo imaging, the nontoxic UC phosphors are very promising for bioimaging applications. In principle, the UC phosphors generally contain less toxic elements. With their photostable luminescence and minimized background autoluminescence by NIR excitation, the UC phosphors show a great potential as alternatives to organic fluorophores and quantum dots for in vivo bioimaging [46, 49]. For example, an intriguing recent development was demonstrated by Chatterjee et al. [54] for in vivo imaging in anesthetized Wistar rats by using 50-nm $NaYF_4$:Yb/Er nanoparticles. The rats were injected with the nanoparticles under the skin in the groin and upper leg regions. Using a combination of simple optical techniques and a 980 nm NIR laser, the nanoparticles can be detected up to 10 mm beneath the skin, far deeper than depths managed through use of conventional equipment based on quantum dots. This method holds the promise of providing a new technique for imaging tissue structures at different depths and for performing minimally invasive detection.

9.4.3 Long Persistent Phosphors

Long persistent phosphors, also called long phosphorescent, long lasting or long afterglow phosphors, are one of the phosphors that have very long afterglow emission. In general, this material emits the emission in the ultraviolet (UV), visible (VIS), or NIR spectral regions for minutes, hours, or even a whole day, after ceasing the energy from excitation source. The emitting of long afterglow is caused by trapped electrons or holes which are produced during the excitation mechanism.

In principle, the luminescence type of this phosphor is phosphorescence in which the lifetime is normally longer than the excited state lifetime and depends on the trap depth and trapping/de-trapping mechanisms. Besides, phosphorescence can be classified by its lifetime into four categories as follows [55]:

(1) Very short persistent phosphorescence (VSPP): the emission has lifetime of the same order of magnitude as the lifetime of excited state. Normally, the emission is no longer than a few milliseconds because of the very shallow traps.
(2) Short persistent phosphorescence (SPP): the emission lasts for seconds and generally becomes prominent to the human eyes. Most common persistent phosphors show SPP after they are exposed to the UV, visible light, or X-rays.

(3) Persistent phosphorescence (PP): the emission lasts for minutes, and it is due to the deep traps in the material. Electrons and holes are held by these traps for some shorts of time, and then they are thermally released from their respective traps.
(4) Long persistent phosphorescence (LPP): the emission demonstrates the lifetimes in order of minutes, hours, or longer.

However, this section focuses only on the long persistent phosphorescence in the aspects of afterglow process, long persistent phosphor structure, and applications with the examples of material.

9.4.3.1 Afterglow Process and Models of Long Persistent Phosphor

The mechanism of LPP can be summarized in the term of energy level diagram (Fig. 9.8) which mainly consists of valence band (VB), conduction band (CB), metastable trapping states for active electrons, and ground and excited states of luminescent center. The process occurs by exciting the electrons from VB to CB via absorption process, and at the same time, charge carriers (electron or hole) are generated and then captured by electron and hole traps which are generally type of electronic defects in materials. Further, these trapped charge carriers will be de-trapped and migrated through CB to the luminescent center. Besides, in this process, there are some energy releases which are called thermal energy, corresponding to thermal de-trapping process. The de-trapping and migration processes may last up to several minutes, hours, or whole day, resulting in a phosphor emitting persistent phosphorescence. The mechanism of LPP is relatively complicated. To easily understand, therefore, in summary, the luminescence mechanism of long persistent phosphors can be divided into four steps [56]:

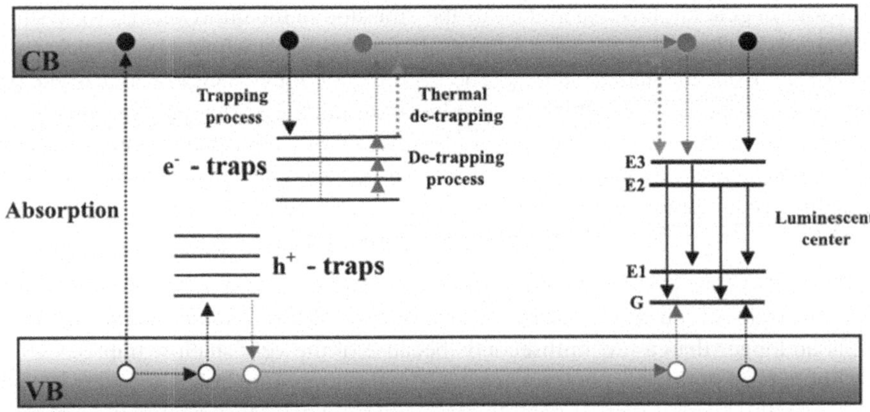

Fig. 9.8 Schematic showing long persistent phosphorescence mechanism. (Adapted from Ref. [57] under Creative Commons Attribution License)

(1) Generating of charge carriers (electron or hole): by the excitation source such as effective optical irradiation (UV, VIS, or NIR light), electronic implantation (e.g., electron beam), and high-energy ray irradiation (e.g., X-ray).
(2) Storage of charge carriers which is used to capture the excited carriers: capacity of the storage depends on the types and numbers of carriers and defects.
(3) Releasing of charge carriers: depending on various external environments such as optical, thermal, or mechanical disturbance.
(4) Recombination process of charge carriers.

It is commonly recognized that LPP outcomes from intrinsic traps or intentionally doped traps or from both. Theoretically, these traps can be separated into two clusters: hole traps that capture holes above the valence band (VB) and electron traps that take electrons below the conduction band (CB). In this schematic, a metastable trapping state for active electron is related to trapping and de-trapping mechanisms which are the important key to understand the principle mechanism of LPP.

Trapping mechanism is usually in accordance with electron excitation and delocalization. After stimulation of an electron from the ground state to excited state, the electron is delivered to an electron trap by the mechanisms which are correlated to the electron delocalization. Under normal environment, at the ground and excited states, electrons are localized around their parent ions. If the electron is excited by appropriate photo-energy and the excited state energy level of the electron is overlapped with the CB, electron delocalization can take place. This process is called photoionization. If the excited state of the electron is below but very close to the CB, electron will have a chance to move up to CB where delocalization can occur by using phonon energy. Hence, this process is called phonon-assisted delocalization. In addition, by taking phonon energy, the electrons are also thermally excited to CB. So, this process could be called thermal ionization. Usually, the thermal ionization is much weaker than photoionization and not strong enough to trap more electrons for producing persistent emission.

De-trapping mechanism is the process that gives rise to the persistent emission. Electron trap usually has a trap depth which is defined as the energy that is caused to the release of trapped electrons. The depth of a trap can be changed from several tenth of an eV to a couple of eV. Trapped electrons can be released from the traps by several de-trapping mechanisms such as phonon-assisted thermal de-trapping, impact ionization, and tunneling. Because de-trapping of a trapped electron usually uses thermal energy (thermal activation), the de-trapping rate A is in accordance with temperature and is given by Eq. (9.15).

$$A = se^{-\Delta E/kT} \tag{9.15}$$

where s is the electron-phonon interaction frequency factor for electron de-trapping, ΔE is the trap depth, k is Boltzmann's constant, and T is absolute temperature. In general, the electron is delocalized through the CB; therefore, the trap depth is typically calculated from the bottom of the CB. The s factor is referred to how many times an electron interacts with a phonon per second at the certain trap level. Once the electron is brought up to a releasing level, it may still be able to turn back to the trap. This process is called re-trapping. The re-trapping is generally described as the possibility of electron falling back to the trap. De-trapping mechanism will be complicated if re-trapping process is considered. For this reason, the first and second order kinetic equation models are proposed and used to explain this complicated mechanism whether the re-trapping process should be considered or not [55, 58, 59].

There is still under argument in the explanation of trapping and de-trapping processes of charge carriers because these processes involve with three important issues, types, concentrations, and depth of effective traps, which are complicated to identify. However, the concrete mechanism of long persistent luminescence has continuously been a hot spot and multifaceted problems. To date, there is no reliable and undoubted explanation, concerning the persistent luminescence mechanism [56]. After reviewing the investigation and experiments of many researchers, there are three models which are widely accepted as the possible models of long persistent phosphorescence, consisting of conduction band-valence band model (CB-VB model), quantum tunneling model, and oxygen vacancy model.

The conduction band-valence band model (CB-VB model) is a simple model for explaining the LPP mechanism in which the energy storage for LPP is mainly concerned. This energy storage is achieved by the localized position of the excited carriers near the CB and VB. When the CB and VB obtain trapped electron or hole, both are free to move until taken by another trap or by a recombination center. This model is based on the participation of CB and VB in the excitation, migration, capturing, and release processes of charge carriers. The key point of this model is all traps should be near the CB or VB.

Quantum tunneling model is proposed based on quantum mechanical phenomena where the pathway of a particle should be the tunnel inside a barrier which the particle is easily movable. This process may take place between the trap positions which are close to excited state levels of the activator. Besides, this model does not take an interest in traps' level, CB, and VB. In general, this model is used to describe the afterglow mechanism associated with deep traps.

Oxygen vacancy model asserts that the persistent phosphorescence of materials is related to the crystal lattice defects. This model involves the oxygen vacancy (V_O) which is formed in the host lattice matrix. The V_O works as the electron trap to capture the electron because the two positively charged units of oxygen vacancy have Coulomb attraction with the electrons in the crystal field; therefore, it can capture the electrons. Also, this model can be explained based on simple luminescence mechanism as follows. In theory, when the electrons are excited to the excited state, part of them will return to the ground state in the form of radiative or non-radiative, while another part will relax to the defect V_O. Further, the electrons in defect V_O may relax to the traps, and then they will be captured. After absorbing more energy, the captured electrons will move off the traps and shift to the excited state of the luminescent center, followed by the same cycle as initial. This mechanism also results in the persistent emission [59].

9.4.3.2 Structure of Long Persistent Phosphor

The long persistent phosphors normally consist of an inorganic crystalline host and dopant ions. Generally, the dopant ions involve two kinds of active centers: luminescent center (activator) and trap. Phosphorescence is physically controlled by the traps concentration and their depth in the host material generated by the deformation of the host or adding some impurities. The emission intensity of persistent phosphorescence is dependent on the concentration of activator and trap, while the depth of trap governs the persistent time.

Host material is very significant for the fabrication of long persistent phosphors. For the activation of their emission, host material plays an important role as trap carrier. Therefore, the trap content is different with the change in host material. Zinc sulfide (ZnS) is the first host material for persistent phosphorescence, but it is very hard to obtain the long persistent emission in this host because it shows very narrow band gaps (2.16 eV), leading to the shallow trap. Later, calcium sulfide (CaS) and the other alkaline-earth sulfides were proposed and found that they emit very strong afterglow emission under visible excitation. Unfortunately, these hosts are chemically unstable because they are easily reacted with moisture, forming hydrogen sulfide gas. Phosphates, rare earth oxides, and oxysulfides are also used as the host for long persistent phosphors. But they cannot be charged by visible light, resulting in the limitation in the applications. Eventually, many researchers have found that alkaline-earth aluminate and silicate hosts exhibit efficient persistent luminescence and become the important host for the reason that many ions can be doped into these hosts and exhibit long persistent emission even without trapping centers because the defects are easily created in these hosts due to the charge compensation and cation disorder. In addition, both contain wide band gap where deep traps can be created. Since then, many long persistent phosphors have been developed by using these hosts, and the most famous hosts for long persistent phosphors are $SrAl_2O_4$, $Sr_4Al_{14}O_{25}$, $CaAl_2O_4$, $Sr_2MgSi_2O_7$, and $Ca_2MgSi_2O_7$ [56].

In the case of dopant ions, transition metals or rare earth elements are typically employed as activator and trap. The activator plays an important role as a persistent luminescent center. Most of the ions, where the 5*d* state or 3*d* state can be populated, are the good candidates for generating persistent emission, for example, rare earth ions which have 5*d* state such as Ce^{3+}, Eu^{2+}, Eu^{3+}, and Tb^{3+}, or transition metals which have 3*d* state such as Mn^{2+}, Mn^{4+}, and Ti^{4+}. These properties are good for long persistent phosphors because the electronic state of these ions is close to the CB of the host which is easy to originate electron delocalization and trapping processes. The Eu^{2+} is the most famous isolated phosphorescent center due to its half shell-filled characteristic, and its emission wavelength from $4f^65d^1$ to $4f^7$ changes from host to host, leading to the good coverage of the spectrum. In the case of trap creation, the second dopant ion will be added to the host and called co-dopant. The adding of co-dopant ion into the host is the process to produce defect-related trapping centers and is one of the most regularly used techniques to make long persistent phosphors. Besides, the persistent lifetime can be increased with the doping of suitable co-dopant. The co-dopants which contain different valences from host cation produce defects because of charge compensation requirements [60]. For example, Y^{3+} and Al^{3+} are doped into CaS:Eu^{2+} to be substituted for Ca^{2+} [61], and Mg^{2+} and Ti^{4+} are doped into Y_2O_2S:Eu^{3+} to replace Y^{3+} [62]. Some ions themselves act as the trapping center when they are co-doped into the host. These ions generally capture either electrons or holes, and then change to metastable ionic states that serve as the trap. Nd^{3+} in $CaAl_2O_4$:Ce^{3+}/Nd^{3+} and Gd^{3+} in $BaAl_2O_4$:Eu^{2+}/Dy^{3+}/Gd^{3+} are typical examples of this type of trap [63]. Ions co-doped into the host not only create additional trapping center but also improve the trapping efficiency. Ions with stronger transition rate can more efficiently pump electrons into the CB, leading to the larger trap populations. An example is Ce^{3+} doped $MgAl_2O_4$ where electrons are pumped into the traps via Ce^{3+} 4*f*-5*d* transitions. This populates traps around 30 times more than those populated through host band gap absorption [64].

The emission color of long persistent phosphors is always important for practical applications. Many phosphors have been developed to obtain the various desired colors, for example, $Sr_4Al_{14}O_{25}$:Eu^{2+},Dy^{3+} and $Ca_2Mg_2Si_2O_7$:Eu^{2+},Dy^{3+} with green afterglow, $CaAl_2O_4$:Eu^{2+},Nd^{3+} and $Sr_3MgSi_2O_8$:Eu^{2+},Dy^{3+} with blue afterglow, Ca_2SiS_4:Eu^{2+},Nd^{3+} with red afterglow, and $CaAl_2Si_2O_8$:Eu^{2+},Mn^{2+},Dy^{3+} with white afterglow. In addition, NIR afterglow activated by transition metals or rare earth elements has been proposed and developed and received much attention because this emission can be applied in the biomedical technologies, especially in vivo imaging applications. The examples of NIR afterglow phosphor are $SrAl_2O_4$: Eu^{2+},Dy^{3+},Nd^{3+} and $CaAl_2O_4$:Eu^{2+},Nd^{3+} [56].

9.4.3.3 Applications

Over the past decade, long persistent phosphors have been extensively studied and applied in various significant applications, for example, optical storages, sensors, detectors, biomedical applications such as in vivo imaging and bio-detection, photocatalysis, or even solar cells. However, the most familiar applications of this material type are luminous indications in the forms of security signs, emergency route signs, safety indication, and indicators of control panels in dark environments or in the night. In this part, an underlying application, luminous indications, and an advance application, in vivo imaging, including some examples of material will be discussed as follows.

The materials for luminous indication applications are typically produced by using phosphorescent material which can glow in the dark for a given duration time after absorption of light by sunlight or electric lamp or bulb light. The long persistent phosphors are one of the candidates of these applications. In general, these materials emit green phosphorescence at night when there is no any light at that place. Nowadays, this application is applied in various fields of emergency signs, for instance, luminous paints in transportation system such as airports and highways and warning signs and escape routes in the buildings. The long persistent phosphors have been widely used in these applications because they are safe, chemically stable, no radiation, nontoxic, and harmless. In addition, the certain brightness of long persistent phosphors is the advantage for emergency lighting because it can be easily detected by human eyes during the emergency situations, leading to the reduction of confusion as well as being energy-saving. The most used materials for luminous indications are Eu-doped aluminates such as $SrAl_2O_4$: Eu^{2+},Dy^{3+} because they are easily produced when compared to other persistent phosphors, and especially they can be produced in powder form as long persistent phosphor which has more advantages in these applications than bulk form [56].

In the field of biomedical application, there is increasing interest in the use of long persistent phosphors as the biological transparent window for in vivo imaging to drive photonic bio-label for tracing cancer cells. A long persistent phosphor is anticipated to open the options of advanced optical bioimaging with high resolution and weak light disturbance for truthfully assessing the structural and functional mechanisms in living cells, tissues, and other complex systems. Various long persistent phosphors have been reported for the recent two decades, but most afterglow wavelengths of conventional long afterglow phosphors are in the visible region which is difficult to penetrate the deeper organic cells or tissues, causing the low optical tracking signal. To overcome this problem, it is well-known that near infrared (NIR) light (wavelength from 650 to 950 nm) can penetrate biological tissues more efficiently than visible light, resulting in the efficient biological transparency windows for in vivo imaging applications. There are many researches on NIR afterglow phosphors, concerning the persistent lifetimes for making good in vivo imaging instruments and targeting the longtime imaging with high sensitivity, for example, $LiGa_5O_8$:Cr^{3+} grafted with PEG-OCH_3 as potential nanoprobe for in vivo imaging with afterglow signal real time more than 1 h [65] and afterglow nanospheres of SiO_2/$SrMgSi_2O_6$:Eu^{2+},Dy^{3+} grafted with PEG-COOH showing afterglow signal real time more than 1 h after peritoneal injection in a live mouse [66].

References

1. C.B. Carter, M.G. Norton, *Ceramic Materials: Science and Engineering* (Springer, New York, 2013)
2. W.A. Weyl, *Coloured glasses* (Society of Glass Technology, Sheffield, 1951)
3. F. Träger, *Springer Handbook of Lasers and Optics* (Springer, Berlin/Heidelberg, 2012)
4. H.J.W. Strutt, *Philos. Mag.* **4**(41), 107 (1871)
5. H.J.W. Strutt, *Philos. Mag.* **4**(41), 447 (1871)
6. F.R.S. Lord Rayleigh, *Philos. Mag.* **5**(12), 81 (1881)
7. F.R.S. Lord Rayleigh, *Philos. Mag.* **5**(47), 375 (1899)
8. S.F. Wang, J. Zhang, D.W. Luo, F. Gu, D.Y. Tang, Z.L. Dong, G.E.B. Tan, W.X. Que, T.S. Zhang, S. Li, L.B. Kong, Prog. Solid State Chem. **41**, 20 (2013)
9. L.B. Kong, Y.Z. Huang, W.X. Que, T.S. Zhang, S. Li, J. Zhang, Z.L. Dong, D.Y. Tang, *Transparent Ceramics* (Springer, Cham, 2015)
10. P.J. Patel, G.A. Gilde, P.G. Dehmer, J.W. McCauley, P. SPIE **4102**, 1 (2000)

11. Y. Imanaka, Y. Suzuki, T. Suzuki, K. Hirao, T. Tsuchiya, H. Nagata, J.S. Cross, *Advanced Ceramic Technologies & Products* (Springer, Tokyo, 2012)
12. ASTM C162-99, *Standard Terminology of Glass and Glass Products* (ASTM International, West Conshohocken, 1999)
13. G.H. Beall, D.A. Duke, J. Mater. Sci. **4**, 340 (1969)
14. G.H. Beall, Annu. Rev. Mater. Sci. **22**, 91 (1992)
15. A. Krell, T. Hutzler, J. Klimke, cfi Ber. DKG. **84**, 6 E41 (2007)
16. L. Pauling, S.B. Hendricks, J. Am. Chem. Soc. **47**, 781 (1925)
17. D.D. Silva, A.R. Boccaccini, Recent Patents Mater. Sci. **1**, 56 (2008)
18. K.M. Kinsman, J. McKittrick, E. Sluzky, K. Hesse, J. Am. Chem. Soc. **77**, 2866 (1994)
19. G. Partridge, Adv. Mater. **2**, 553 (1990)
20. R. Royon, J. Lhermite, L. Sarger, E. Cormier, Opt. Express **21**, 13818 (2013)
21. N. Soleimani, M. Ponting, E. Gebremichael, A. Ribuot, G. Maxwell, J. Cryst. Growth **393**, 18 (2014)
22. Y. Zorenko, V. Gorbenko, T. Zorenko, A. Banaszak, L. Mosinska, K. Paprocki, Y. Zhydachevskii, A. Suchocki, P. Bilski, A. Twardak, A. Fedorov, Radiat. Meas. **90**, 308 (2016)
23. A.G. Petrosyan, H.R. Asatryan, K.L. Hovhannesyan, M.V. Derdzyan, S.P. Feofilov, A.V. Eganyan, R.S. Sargsyan, Mater. Chem. Phys. **185**, 39 (2017)
24. R. Boulesteix, A. Maître, J.-F. Baumard, Y. Rabinovitch, Mater. Lett. **64**, 1854 (2010)
25. R. Boulesteix, A. Maître, L. Chrétien, Y. Rabinovitch, C. Sallé, J. Am. Chem. Soc. **96**, 1724 (2013)
26. R. L. Coble, Transparent alumina and method of preparation, U.S. Patent No. 3026210 A, 1962
27. J. Cheng, D. Agrawal, Y. Zhang, R. Roy, Mater. Lett. **56**, 587 (2002)
28. K.Q. Dang, S. Takei, M. Kawahara, M. Nanko, Ceram. Int. **37**, 957 (2011)
29. M. Nanko, K.Q. Dang, Adv. Appl. Ceram. **113**, 80 (2014)
30. L. Lallemant, G. Fantozzi, V. Garnier, G. Bonnefont, J. Eur. Ceram. Soc. **32**, 2909 (2012)
31. N. Roussel, L. Lallemant, J.-Y. Chane-Ching, S. Guillemet-Fristch, B. Durand, V. Garnier, G. Bonnefont, G. Fantozzi, L. Bonneau, S. Trombert, D. Garcia-Gutierrez, J. Am. Ceram. Soc. **96**, 1039 (2013)
32. R. Johnson, P. Biswas, P. Ramavath, R.S. Kumar, G. Padmanabham, T. Indian Ceram, Society **71**, 73 (2012)
33. A.F. Shchurov, V.A. Perevoshchikov, T.A. Gracheva, N.D. Malygin, D.N. Shevarenkov, E.M. Gavrishchuk, V.B. Ikonnikov, E.V. Yashina, Inorg. Mater. **40**, 96 (2004)
34. G. Blasse, B.C. Grabmaier, *Luminescent Materials* (Springer, Berlin/Heidelberg, 1994)
35. S.W.S. Mckeever, *Thermoluminescence of Solids* (Cambridge University, Cambridge, 1985)
36. D.R. Vij, *Luminescence of Solids* (Springer, New York, 1998)
37. R.C. Ropp, *Luminescence and the solid state*, 2nd edn. (Elsevier, Amsterdam, 2004)
38. W.M. Yen, M.J. Weber, *Inorganic Phosphors, Compositions, Preparations and Optical Properties* (CRC Press, Boca Raton, 2004)
39. G. Liu, B. Jacquier, *Spectroscopic Properties of Rare Earths in Optical Materials* (Springer, Berlin/Heidelberg, 2005)
40. X. Huang, S. Han, W. Huang, X. Liu, Chem. Soc. Rev. **42**, 173 (2013)
41. D.L. Dexter, Phys. Rev. **108**, 630 (1957)
42. Q.Y. Zhang, X.Y. Huang, Prog. Mater. Sci. **55**, 353 (2010)
43. K. Grzeszkiewicz, L. Marciniak, W. Stręk, D. Hreniak, J. Lumin. **177**, 172 (2016)
44. A. Verma, S.K. Sharma, Ceram. Int. **43**, 8879 (2017)
45. F. Auzel, Chem. Rev. **104**, 139 (2004)
46. F. Wang, X. Liu, Chem. Soc. Rev. **38**, 976 (2009)
47. M. Pollnau, D.R. Gamelin, S.R. Lüthi, H.U. Güdel, Phys. Rev. B **61**, 3337 (2000)
48. J.F. Suyver, A. Aebischer, S. García-Revilla, P. Gerner, H.U. Güdel, Phys. Rev. B **71**, 125123 (2005)
49. J. Zhou, Q. Liu, W. Feng, Y. Sun, F. Li, Chem. Rev. **115**, 395 (2015)

50. I. Hyppänen, J. Hölsä, J. Kankare, M. Lastusaari, L. Pihlgren, J. Nanomater, *Article ID 16391*, vol 1 (2007)
51. H. Lian, Z. Hou, M. Shang, D. Geng, Y. Zhang, J. Lin, *Energy* **57**, 270 (2013)
52. J. Hampl, M. Hall, N.A. Mufti, Y.M. Yao, D.B. MacQueen, W.H. Wright, D.E. Cooper, Anal. Biochem. **288**, 176 (2001)
53. R.S. Niedbala, H. Feindt, K. Kardos, T. Vail, J. Burton, B. Bielska, S. Li, D. Milunic, P. Bourdelle, R. Vallejo, Anal. Biochem. **293**, 22 (2001)
54. D.K. Chatterjee, A.J. Rufaihah, Y. Zhang, Biomaterials **29**, 937 (2008)
55. W.M. Yen, S. Shionoya, H. Yamamoto, *Phosphor Handbook*, 2nd edn. (CRC Press, Boca Raton, 1995)
56. Y. Li, M. Gecevicius, J. Qiu, Chem. Soc. Rev. **45**, 2090 (2016)
57. J. Thirumalai, *Luminescence – An Outlook on the Phenomena and their Applications* (InTech, 2016), p. 17
58. G. Chen, Z. Xu, J. Appl. Phys. **106**, 123707 (2009)
59. R.-S. Liu, *Phosphors, Up Conversion Nano Particles, Quantum Dots and Their Applications*, vol 2 (Springer, Berlin/Heidelberg, 2016), p. 167
60. D. Jia, X.J. Wang, W.M. Yen, Phys. Rev. B **69**, 235113 (2004)
61. D. Jia, J. Zhu, B. Wu, J. Electrochem. Soc. **147**, 386 (2000)
62. X. Wang, Z. Zhang, Z. Tang, Y. Lin, Mater. Chem. Phys. **80**, 1 (2003)
63. H. Yamamoto, T. Matsuzawa, J. Lumin. **72**, 287 (1997)
64. D. Jia, W.M. Yen, J. Lumin. **101**, 115 (2003)
65. X. Fu, C. Liu, J. Shi, H. Man, J. Xu, H. Zhang, Opt. Mater. **36**, 1792 (2014)
66. Z.-J. Li, H.-W. Zhang, M. Sun, J.-S. Shen, H.-X. Fu, J. Mater. Chem. **22**, 24713 (2012)

Chapter 10
Organic-Inorganic Hybrid Materials

Yoshiyuki Sugahara

Abstract Organic-inorganic hybrid materials have developed rapidly to become an established part of ceramic science and technology. In this chapter, among preparation methods of organic-inorganic hybrid materials, the sol-gel process, intercalation, and surface modification, which have a close relationship with ceramic materials, are described with an emphasis on their synthetic aspects and their applications. To begin with, classification of organic-inorganic hybrid materials based on the type of their interfacial chemical bonds is introduced. Following a brief definition of the sol-gel process, the basic chemistry of the sol-gel process is explained. Various preparation methods for sol-gel-derived organic-inorganic hybrid materials, including bridged silsesquioxane and mesostructured materials, are then described, and representative applications of sol-gel-derived organic-inorganic hybrid materials are briefly discussed. For intercalation, which is characteristic of layered materials, the basic concept is introduced first, after which the applicable reaction mechanisms for intercalation are demonstrated with the grafting reaction, which is related to intercalation. Representative applications of intercalation compounds are then exhibited with applications of nanosheets obtained via the exfoliation of layered materials. The process of surface modification of metal oxide nanoparticles and the application of the resulting surface-modified metal oxide nanoparticles are then demonstrated.

Keywords Organic-inorganic hybrid materials · Metal oxide nanomaterials · Polymers · Sol-gel process · Intercalation · Surface modification

---

Y. Sugahara (✉)
Department of Applied Chemistry, School of Advanced Science and Engineering, Waseda University, Tokyo, Japan

Kagami Memorial Research Institute for Materials Science and Technology, Waseda University, Tokyo, Japan
e-mail: ys6546@waseda.jp

J. Hojo (ed.), *Materials Chemistry of Ceramics*,
https://doi.org/10.1007/978-981-13-9935-0_10

10.1 Organic-Inorganic Hybrid Materials: Introduction [1, 2]

Organic-inorganic hybrid materials have been attracting increasing attention as next-generation materials, since they could exhibit the advantages of both inorganic and organic components. To optimize the properties of organic-inorganic hybrid materials, it is important to control the domain sizes and shapes of the components and chemical bonds at their interfaces, in addition to their compositions and structures. The selection of starting materials and processes is thus important for obtaining desirable macroscopic structures. In this chapter, the sol-gel process, intercalation, and surface modification are selected among various processes for preparing organic-inorganic hybrid materials, since they can provide organic-inorganic hybrid materials where ceramic materials play an important role. Their characteristics and some examples are presented in the following sections.

Among various classifications of organic-inorganic hybrid materials, the classification using interfacial bonds is frequently employed [3]. The advantage of this classification is that it can be applied to organic-inorganic hybrid materials prepared through all the processes.

Class I: Organic-inorganic hybrid materials bearing relatively weak chemical bonds, such as the van der Waals force, hydrogen bonds, and ionic bonds.

Class II: Organic-inorganic hybrid materials bearing relatively strong chemical bonds, covalent and ion o-covalent bonds.

In this chapter, typical processes for producing organic-inorganic hybrid materials in which ceramic materials play key roles are presented: the sol-gel process, inorganic network formation process materials; intercalation compounds and nanosheet-related materials, intercalation reactions, which are host-guest reactions, and related reactions, grafting reactions and exfoliation; and surface modification of nanoparticles are all described.

10.2 Sol-Gel Process [4–8]

10.2.1 Overview of the Sol-Gel Process

The definition of the sol-gel process by IUPAC is as follows: “a process through which a network is formed from solution by a progressive change of liquid precursor(s) into a sol, to a gel, and in most cases finally to a dry network.” [9] The term “sol-gel process” has actually been employed for processes which proceed via formation of no clear sol and gel. Thus, this term has been employed for the conversion of precursors into solid materials, and the precursors are commonly metal alkoxides. Recently developed non-hydrolytic sol-gel process provides typical examples of processes which proceed without the formation of sol and gel [10, 11].

Metal alkoxides are generally expressed as $M(OR)_n$, where M is metal or metalloid, and R is an alkyl or alkenyl group. In extended usage, R could be an organic group with atoms other than hydrogen and carbon, such as oxygen. A large number of metals and metalloids are known to form metal alkoxides.

Some metal alkoxides can form oligomeric or polymeric structures in solvents, and in some cases the alkoxides are insoluble in common organic solvents. The introduction of a bulky group as an R group is known to be an effective means of increasing their solubility. Since metal oxides can be prepared from metal alkoxides, they can be regarded as monomers of metal oxides. Metal alkoxides are not classified as organometallic compounds but as metalorganic compounds. There are a few exceptions, such as organoalkoxysilanes, $R'_xSi(OR)_{4-x}$, where Si-C covalent bonds are present. Since metal alkoxides are hydrolysable molecules, their handling is similar to that of organometallic compounds, which are frequently moisture-sensitive. A typical organometallic-chemistry-type handling procedure involves purification by vacuum distillation.

Although various organic solvents can be employed for dissolving metal alkoxides, alcohols, in particular parent alcohols (alcohols whose alkyl group is the same as the R group in $M(OR)_n$), are utilized. When alcohols other than the parent alcohol are used, reactions called "alcohol-exchange reactions" proceed:

$$\equiv M - OR + R'OH \quad \rightarrow \quad \equiv M - OR' + ROH \tag{10.1}$$

In a hydrolytic sol-gel process, which is commonly employed, hydrolysis of an alkoxy group leads to the formation of a hydroxyl group and triggers a condensation reaction. Such reactions can be expressed as follows for tetraalkoxysilane, $Si(OR)_4$:

[hydrolysis]

$$\equiv Si - OR + H_2O \rightarrow \equiv Si - OH + ROH \tag{10.2}$$

[condensation]

$$\equiv Si - OH + HO - Si \equiv \rightarrow \equiv Si - O - Si \equiv + H_2O \tag{10.3}$$

$$\equiv Si - OH + RO - Si \equiv \rightarrow \equiv Si - O - Si \equiv + ROH \tag{10.4}$$

Since tetraalkoxysilanes are relatively less reactive toward hydrolysis, a catalyst, generally an acid or a base, is added to the system. In an acid-catalyzed hydrolysis, the hydrolysis is initiated by the addition of a proton to an oxygen atom in tetraalkoxysilane, while a direct nucleophilic attack by an OH^- ion triggers the reaction in a base-catalyzed hydrolysis. The hydrolysis rate depends on various factors, including the catalyst concentration and the size of R group. On the other hand, the reactivities of organoalkoxysilanes, $R'_xSi(OR)_{4-x}$, depend on both the R and R′ groups. The inductive effects of R′ groups influence their hydrolysis rates, since they affect the electron density of silicon. Thus, with an increase in x, the hydrolysis rate increases in an acidic system, while the hydrolysis rate decreases in a basic system. The steric effects of both R and R′ groups also change their hydrolysis rates.

The hydrolysis rates of transition metal alkoxides are much faster compared to those of the tetraalkoxysilanes and organoalkoxysilanes [12]. And the hydrolysis rate strongly depends on the metal. One factor is lower electronegativities of the transition metals (the electronegativity of silicon is 2.5 and those of many transition metals are less than 2). Another important factor is larger coordination numbers of the transition metals, coordination numbers larger than four. The coordinative unsaturation of monomeric metal alkoxides also causes additional intermolecular coordination of oxygen atoms to form oligomeric or polymeric structures (Fig. 10.1).

To reduce the reactivities of transition metal alkoxides, modification of transition metal alkoxides is frequently applied. A typical example is modification with β-diketones (HZ). Since β-diketones undergo keto-enol tautomerism, their enol forms react to form β-diketonates:

$$M(OR)_n + HZ \rightarrow MZ(OR)_{n-1} + ROH \tag{10.5}$$

In the resulting molecule, the oxygen atom in the carbonyl group (C=O) coordinates to the metal additionally to reduce the reactivity of the molecule, leading to suppression of the formation of precipitation during the sol-gel process (Fig. 10.2).

Non-hydrolytic processes have been developed in recent decades [10, 11]. Since the presence of water favors a hydrolytic sol-gel process, a non-hydrolytic process is

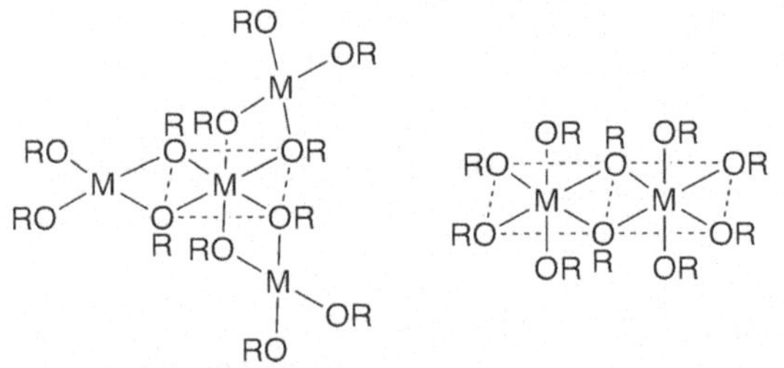

Fig. 10.1 Typical polymeric structures of metal alkoxide

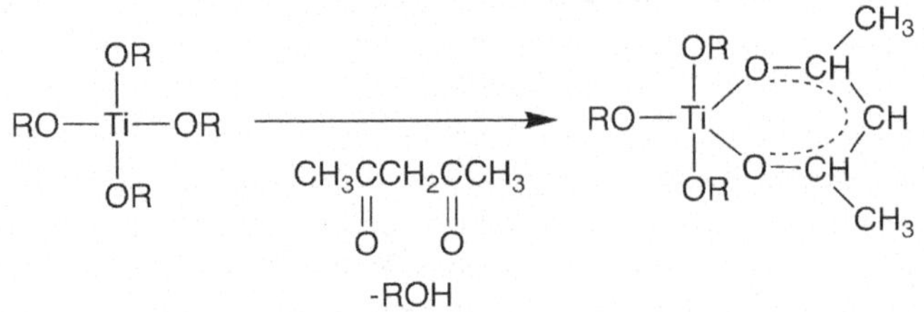

Fig. 10.2 Modification of titanium alkoxide with acetylacetone

conducted in non-aqueous systems. These reactions contain a condensation step, which involves functional groups containing oxygen. Non-hydrolytic processes are particularly suitable for the preparation of organic-inorganic hybrid materials in systems containing moisture-sensitive compounds and those for which controllable condensation is advantageous [13].

10.2.2 Preparation of Organic-Inorganic Hybrid Materials via the Sol-Gel Process

Since thermal treatment at high temperatures is not required for the sol-gel process, the sol-gel process is highly suitable for the preparation of organic-inorganic hybrid materials. Various approaches have therefore been reported to date, with a large number of combination of organic and inorganic components. Typical examples are presented below.

(a) Organic-inorganic hybrid materials prepared via incorporation of organic ions or molecules upon hydrolysis and condensation of metal alkoxides [14, 15]

It is possible to prepare organic-inorganic hybrids containing organic ions or molecules upon hydrolysis and condensation of metal alkoxides. An organic species soluble in a system can be easily incorporated into a final xerogel. Since the resulting organic-inorganic hybrids are class-I hybrid materials, durability can be a problem.

(b) Organic-inorganic hybrids prepared via co-hydrolysis of organoalkoxysilane and tetra-functional alkylalkoxysilane [15, 16]

Since the Si-C bonds are stable with respect to hydrolysis, organoalkoxysilanes are suitable starting materials for class-II hybrid materials. It should be noted that most other metal-carbon bonds are not stable with respect to hydrolysis. It should also be noted that current organosilicon chemistry can provide a variety of commercially available organoalkoxysilanes and also allow us to prepare target organoalkoxysilanes.

Co-hydrolysis of a di-functional or tri-functional organoalkoxysilane ($R'_x Si(OR)_{4-x}$, $x = 1,2$) with tetraalkoxysilane ($Si(OR)_4$) is generally employed to prepare organic-inorganic hybrid materials, which have been called ORMOCERs (organically modified ceramics) or ORMOSILs (organically modified silicate). The hydrolysis and condensation rates of organoalkoxysilanes are different from those of tetraalkoxysilanes, and the hydrolysis and condensation rates of organoalkoxysilanes depend strongly on the R and R′ groups in $R'_x Si(OR)_{4-x}$, as discussed above. Thus, when R is a small organic group such as a methyl group, a relatively homogeneous structure can be obtained (Fig. 10.3), while a segregated structure can be obtained with a bulky R group.

By employing an organoalkoxysilane with a polymerizable group, another organic polymeric network can be formed in addition to an inorganic siloxane network, leading to a relatively homogeneous structure. Structures of this type in

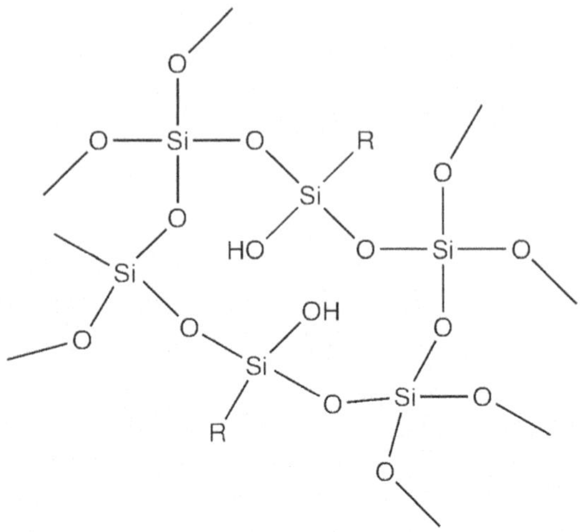

Fig. 10.3 Schematic drawings of ORMOCERs (organically modified ceramics)

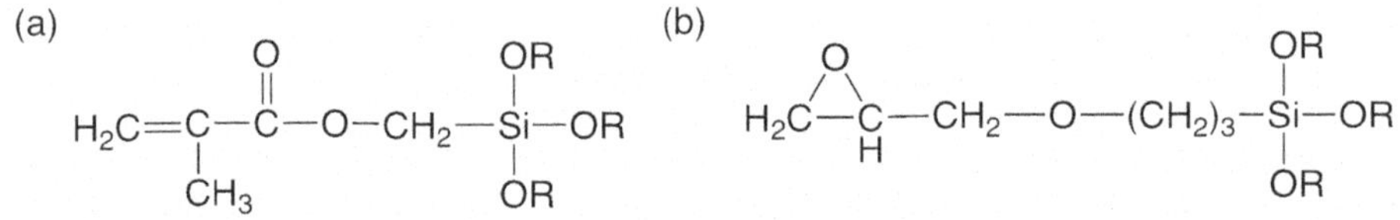

Fig. 10.4 Examples for organoalkoxysilanes bearing polymerizable functional groups; (**a**) methacryloxymethyltrialkoxysilane and (**b**) (3-glycidoxypropyl)trialkoxysilane

which an organic network and a metaloxane network are penetrating each other are classified to so-called "interpenetrating polymer networks" (IPNs). Typical organoalkoxysilanes are methacryloxymethyltrialkoxysilane and (3-glycidoxypropyl) trialkoxysilane (Fig. 10.4).

(c) Organic-inorganic hybrids prepared from organic polymers and metal alkoxides [17]

When an organic-inorganic hybrid is prepared from an organic polymer and a metal alkoxide via the sol-gel process, the affinity between the organic polymer and a metaloxane network formed from the metal alkoxide plays a crucial role, and the two components could be easily segregated. Thus, preparation should be carefully designed to achieve preparation of homogeneous organic-inorganic hybrids.

A homogeneous organic-inorganic hybrid material can be obtained by selecting an appropriate polymer structure. Various hydroxyl-terminated and trialkoxysilyl-functionalized polymers have been employed with a metal alkoxide or tetraalkoxysilane to prepare homogeneous organic-inorganic hybrid materials [18]. Polymeric structures employed for homogeneous organic-inorganic hybrid materials include polydimethylsiloxane (PDMS), poly(tetramethylene oxide) (PMTO), poly(methyl methacrylate) (PMMA), and polyoxazoline. When tetraalkoxysilane and silanol-terminated PDMS are used, the resulting organic-inorganic hybrids are also called ORMOSILs. For a homogeneous silica-polyoxazoline hybrid material, a structural model in which hydrogen bonding between polyoxazoline and a siloxane network plays an important role was proposed (Fig. 10.5) [19].

Another approach involves a polymerizable molecule released upon hydrolysis of tetraalkoxysilane. Typical metal alkoxides are shown in Fig. 10.6 [20]. In a standard sol-gel process, alcohol is released upon hydrolysis of a metal alkoxide, and the released alcohol molecules, which are normally in the form of a volatile liquid, cause shrinkage upon drying. The molecules released from this type of tetraalkoxysilane, on the other hand, can be polymerized to form a polymeric matrix.

(d) Bridged-silsesquioxane-type organic-inorganic hybrids prepared from precursors bearing multiple trialkoxysilyl groups [18, 21–25]

Another type of organic-inorganic hybrids, bridged-silsesquioxanes, has been developed. Bridged-silsesquioxanes are prepared via the sol-gel process from

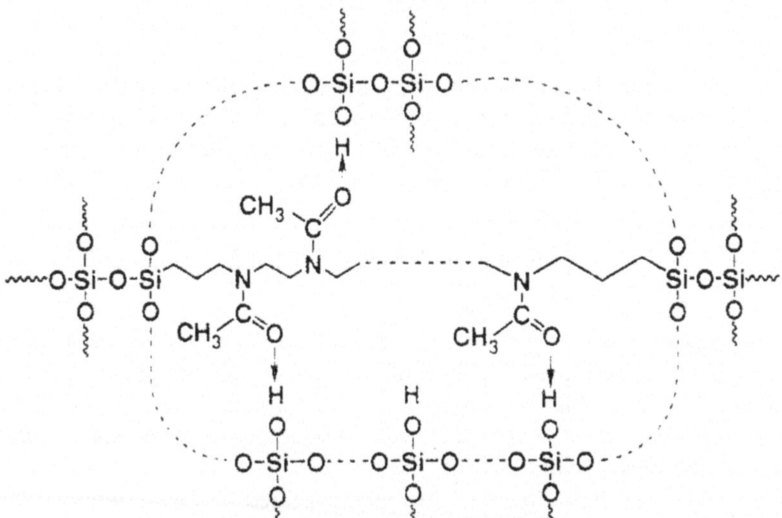

Fig. 10.5 Proposed structures of a silica-polyoxazoline hybrid material. (Reprinted with permission from Ref. [19]. Copyright (1993) American Chemical Society [19])

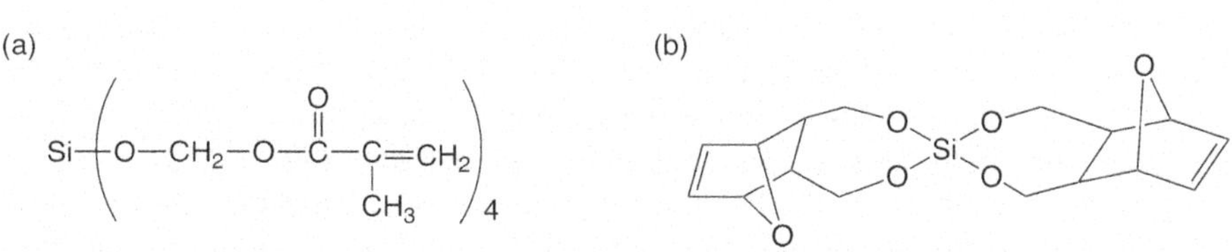

Fig. 10.6 Tetraalkoxysilanes bearing polymerizable groups in alkoxy groups; (**a**) methacryloxymethyltrialkoxysilane and (**b**) (3-glycidoxypropyl)trialkoxysilane

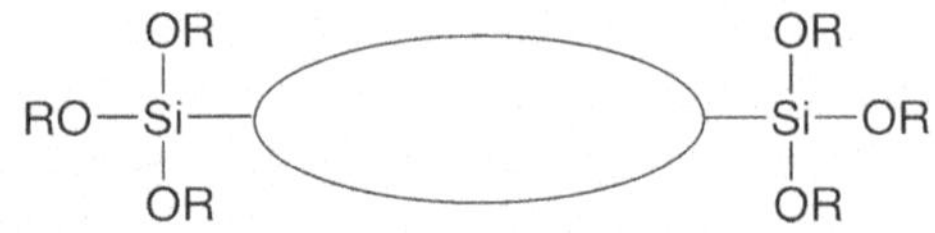

Fig. 10.7 Schematic representation of a precursor for bridged polysilsesquioxane

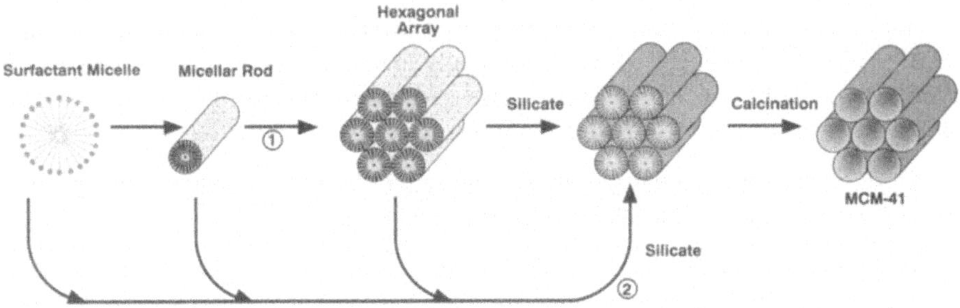

Fig. 10.8 Proposed mechanism for mesoporous silica. (Reprinted with permission from Ref. [38]. Copyright (1992) American Chemical Society [38])

precursors bearing multiple trialkoxysilyl groups, $R[Si(OR)_3]_n$ typically with $n = 2$ (Fig. 10.7). The advantage of this type of precursor is facile gelation without the addition of another metal alkoxide, and relatively homogeneous organic-inorganic hybrids can be obtained. Thus, enormous effort has been directed to developing bridged silsesquioxanes in recent years.

(e) Organic-inorganic hybrids prepared using colloid and supramolecular chemistry [26–29]

A combination of the sol-gel process with colloid chemistry has been applied to the preparation of nanomaterials in confined media and biomimetic silica-based membranes [30, 31]. Another outcome is the formation of microscale helical organic-inorganic hybrid materials using bridged-type precursors with two trialkoxysilyl groups, $R[Si(OR)_3]_2$ [32, 33].

A robust new research direction, however, appeared following the discovery of the ability of the liquid crystal templating process [34–36] (Fig. 10.8) to provide mesostructured materials, which can be converted into mesoporous materials via the removal of templates [37, 38]. In the standard preparation of mesoporous silica, which is typical materials prepared via the liquid crystal templating process, the electrostatic interaction between positively charged templates, which is formed by the self-assembly of cationic surfactant, and anionic silicate species plays a crucial role in the formation of mesostructured class-I hybrid materials. Hydrolysis of precursors bearing Si-C bonds in the presence of liquid crystal templates leads to the formation of mesostructured class-II hybrid materials [39, 40]. The use of bridged-type precursors, $R[Si(OR)_3]_2$, in the liquid crystal templating process provides class-II mesoporous hybrid materials called periodic mesoporous organosilica (PMO) [41].

Sol-gel-derived silica-based mesoporous materials (mesoporous silicas) are capable of surface modification using the reactivity of surface silanols by silylation, where surface silanol groups react with silylation reagents, typically $XSiR_3$, where $X = Cl$ or OR', and the overall reaction can be expressed as shown below [42]:

$$\equiv SiOH + XSi\equiv \quad \rightarrow \quad \equiv SiOSi\equiv + HX \qquad (10.6)$$

10.2.3 *Applications of Organic-Inorganic Hybrid Materials via the Sol-Gel Process [15, 43]*

A variety of applications have been proposed and studied for sol-gel-derived materials. A few notable applications are described below.

(a) Photofunctional organic-inorganic hybrids [44–47]

Since transparent and robust materials can be prepared by the sol-gel process, various photofunctional organic-inorganic hybrids have been prepared. When a functional dye or a luminescent metal complex is added to the sol-gel system, for example, these guest species can be immobilized.

Organically modified siloxane networks obtained by the co-hydrolysis process can also be employed as hosts for encapsulating functional molecules and ions. For hydrophobic dyes, in particular, standard sol-gel matrices derived through hydrolysis of tetraalkoxysilanes are not appropriate matrices. In addition, since some functional molecules and ions are sensitive to their environments, their photochemical and photophysical behavior might depend on the compositions of organic-inorganic hybrids [48]. Thus, organically modified siloxane networks are attractive matrices. Typical example is spyrooxazine in sol-gel matrices derived from a vinyltriethoxysilane-tetraethoxysilane (VTEOS-TEOS) system. Spyrooxazine was present in a gel derived from vinyltriethoxysilane alone, while merocyanine was present in a gel prepared from a TEOS-rich system (Fig. 10.9) [49].

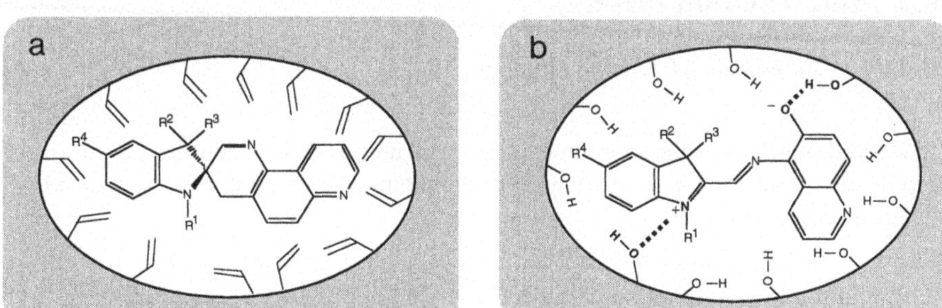

Fig. 10.9 Two forms of spirooxazine in hybrid gels; (**a**) spirooxazine in a gel derived from vinyltriethoxysilane and (**b**) a twitter ion form (merocyanine) in a gel from a vinyltriethoxysilane-tetraethoxysilane system. (Reprinted with permission from Ref. [49]. Copyright (1996) American Chemical Society [49])

Coupling reactions and condensation reactions can be employed for binding organic functional groups covalently to silicon to prepare functional organoalkoxysilanes. A typical example is second harmonic generation (SHG) materials, since simple encapsulation of SHG dyes results in instability of dye alignment caused by poling. Thus, SHG dyes with trialkoxysilyl groups were synthesized and employed to prepare class-II SHG hybrid materials via a co-hydrolysis process [50].

Surface modification of sol-gel-derived mesoporous silica with surface modification agents bearing photofunctional groups led to class-II hybrid materials [51–53]. The use of photofunctional groups as spacers for the bridged precursors leads to the formation of various photofunctional PMOs as Class-II hybrid materials [29].

Another topic is immobilization of luminescent metal complexes, especially lanthanide complexes, which exhibit excellent luminescent properties [54]. When a soluble metal complex is added to a sol-gel system, a lanthanide complex can be easily encapsulated in a sol-gel matrix. Introducing trialkoxysilyl groups to a ligand enables a lanthanide complex with multiple trialkoxysilyl groups to be prepared, and a bridged-silsesquioxane-type material can be obtained by the sol-gel process.

(b) Organic-inorganic hybrids encapsulating biological materials [55–59]

Since the sol-gel process is suitable for the immobilization of biological materials, a large number of studies have been reported on immobilized biological materials ranging from enzymes to whole protozoa.

Among biological materials, encapsulation of enzymes is attracting a particularly enormous amount of attention; encapsulation of enzymes is important for the development of biocatalytic processes, and additional advantages such as better thermal stability are expected. In a typical example, activities of phosphatases encapsulated in sol-gel matrices were maintained at a pH where bare phosphatases lost their activities [60].

10.3 Intercalation [61–67]

10.3.1 Overview of Intercalation

For layered materials, intercalation is defined as an insertion reaction of a guest ion or molecule between layers. A standard intercalation reaction can be expressed by the following equation:

$$xG + \square_x[H] \rightleftarrows G_x[H] \tag{10.7}$$

where [H] is the host substance, G is the guest species, and $\square$ represents an interlayer vacant site which can accommodate the guest species. The product, $G_x[H]$, is called an intercalation compound. It should be noted that intercalation reactions are reversible.

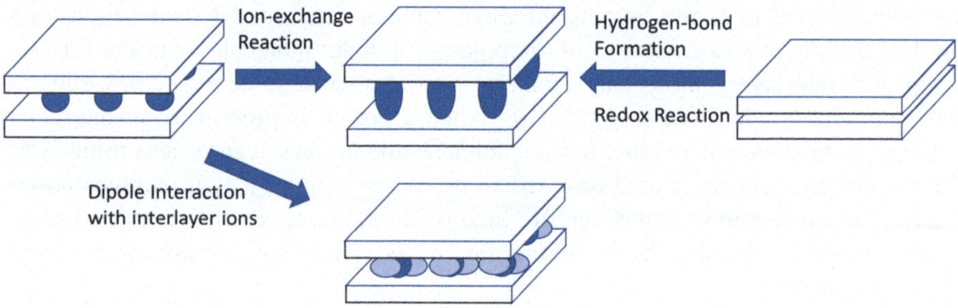

Fig. 10.10 Typical mechanisms of intercalation reactions

In this section, intercalation reactions of metal oxide-based layered materials are described, and grafting reactions and exfoliation, the latter of which is currently attracting increasing attention, are also briefly described. Brief descriptions of other typical layered materials such as graphite and iron oxychloride are also included.

10.3.2 Intercalation Mechanism

(a) Ion exchange

Figure 10.10 shows the concept of intercalation via ion exchange. When inorganic ions are present in the interlayer space, organic ions can be introduced by ion exchange. Typical host substances for cation exchange are montmorillonite, a typical aluminosilicate clay, layered transition metal oxometalates, such as layered titanates and niobates, and layered zirconium phosphates; *n*-alkylammonium ions are typical guest species. Anion-exchangeable layered materials are limited; a typical example is layered double hydroxides (LDHs).

(b) Hydrogen bonding

In some neutral layered materials bearing surface hydroxyls, guest species can be present between the layers due to the formation of hydrogen bonds (Fig. 10.10). A typical example is kaolinite, a layered aluminosilicate clay. In addition to polar organic molecules such as dimethyl sulfoxide, carboxylate salts, such as potassium acetate, can be present between the layers of kaolinite.

(c) Redox reaction

When a host substance is capable for redox reactions, intercalation involving a redox reaction could proceed (Fig. 10.10). A typical example is layered transition metal dichalcogenides and graphite.

(d) Interaction with interlayer ions

Some layered materials bearing interlayer cations accommodate water molecules in their interlayer space because of the polarity of water molecules. Similar interactions with interlayer cations cause intercalation of polar organic molecules, such as alcohols and acrylonitrile (Fig. 10.10). When a proton is present as an interlayer cation, on the contrary, organic bases, such as *n*-alkylamines, can be accommodated in the interlayer space by acid-base reactions. When relatively hydrophobic cations such as tetramethyammonium ions are incorporated, moreover, the resulting organically pillared relatively hydrophobic interlayer space can accommodate small organic molecules.

(e) Grafting reaction

If reactive groups are present on the interlayer surfaces, grafting reactions could proceed. It should be noted that these grafting reactions are not truly categorized as intercalation reactions, which should be reversible. A well-known host is iron oxychloride, FeOCl, where surface Fe-Cl bonds undergo grafting reactions with alcohols and sodium alkoxides:

$$\mathrm{FeCl + ROH \rightarrow FeOR + HCl} \quad (10.8)$$

$$\mathrm{FeCl + RONa \rightarrow FeOR + NaCl} \quad (10.9)$$

Among metal oxide-based layered materials, layered crystalline silicic acids, such as magadiite and kenyaite, are well-known examples [68, 69]. The standard technique for grafting reactions is silylation (Fig. 10.11) [70, 71]. For transition metal oxometalates, silylation can also be applied [72]. Phosphorous coupling agents can be applied as well, moreover, to form stable M-O-P bonds [73].

Oxidized graphite, graphite oxide, also possesses reactive groups and undergoes grafting reactions [74].

10.3.3 Preparation of Organic-Inorganic Hybrid Materials via Intercalation Reactions and Exfoliation

A variety of combinations of host substances and guest species have been employed for intercalation reactions, and appropriate synthesis strategies have been applied according to the combination and target application. Exfoliation has been employed, on the contrary, to prepare nanosheets. Typical examples of intercalation reactions and exfoliation are shown below.

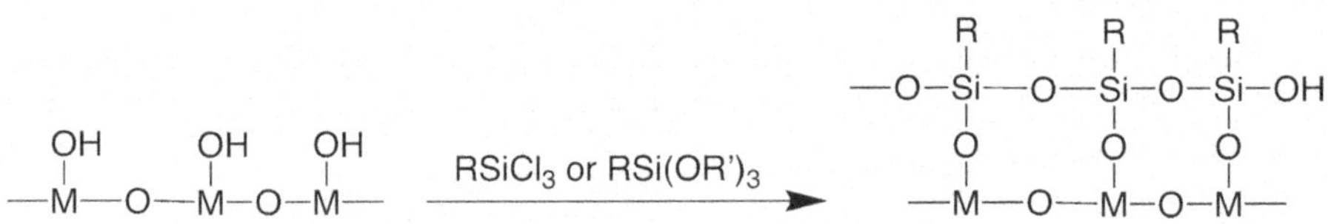

Fig. 10.11 Silylation reaction of a layered material

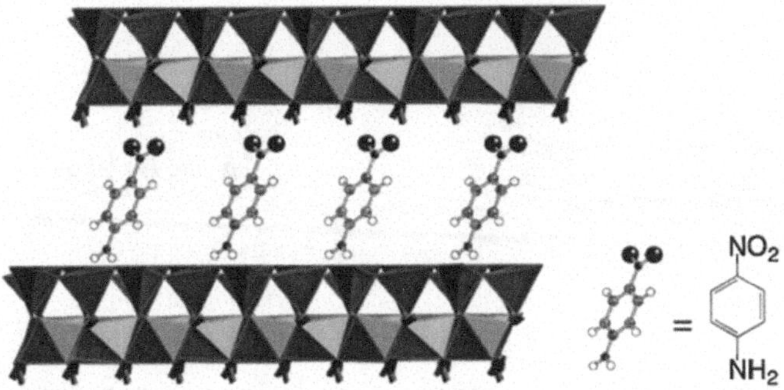

Fig. 10.12 Intercalation of *p*-nitroaniline into kaolinite. (Reprinted with permission from Ref. [79]. Copyright (2001) American Chemical Society [79])

(a) Immobilization of functional ions and molecules [75, 76]

Intercalation of functional organic molecules and ions leads to organic-inorganic hybrid materials in which organic molecules or ions are homogeneously dispersed in the interlayer space. Cationic functional organic compounds such as rhodamine 6G can be immobilized especially easily in the interlayer space [77]. One strategy for neutral functional organic compounds is their accommodation in porous structures prepared by the intercalation of organic cations: the vapor-phase intercalation of *p*-nitroaniline, a typical second harmonic generation dye, under an electric field to achieve its noncentrosymmetric alignment [78]. The alignment of *p*-nitroaniline was also achieved by intercalation in kaolinite, whose interlayer environment was asymmetrical with a silicate layer and an aluminol layer facing each other (Fig. 10.12) [79].

The immobilization of metal complexes, which act as homogeneous catalysts, in the interlayer space provides unique solid catalysts. For example, immobilization of a Ziegler–Natta polymerization catalyst in fluorinated mica-type silicate and swelled synthetic hectorite gave solid catalysts active for polymerization of propylene with a narrow molecular mass distribution [80].

Encapsulation of enzymes can be achieved by intercalation, and the resulting materials can be employed as biocatalysts [81]. A typical study involved the intercalation of several enzymes in the interlayer space of α-zirconium phosphate, and the resulting intercalation compounds exhibit catalytic activities [82].

(b) Interlayer design for selective adsorption [67, 83, 84]

Since host substances are generally crystalline, the reactive sites are expected to be ordered on the interlayer surfaces. Such regularity can be utilized for constructing interlayer environments for selective adsorption. Selective adsorption by layered materials was first reported for vanadyl alkylphosphonates [85] and was also achieved by magadiite modified with octyl groups via a surface grafting reaction. Unique interlayer environments can be achieved by immobilizing optically active iron(II) tris(1,10-phenanthroline); montmorillonite bearing a single enantiomer in

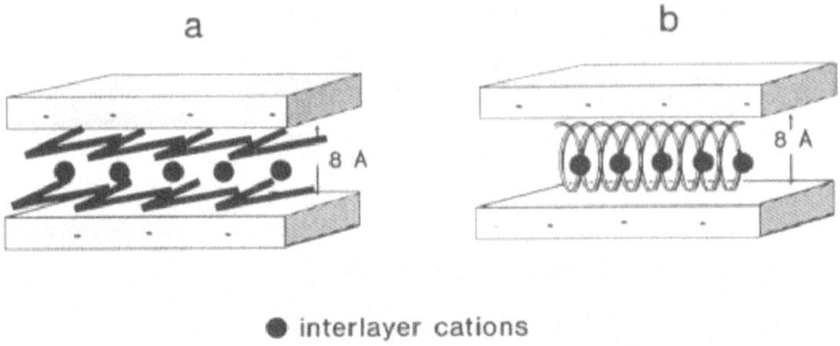

Fig. 10.13 Schematic representation of PEO intercalation models in phyllosilicates: (**a**) double-layer planar zigzag disposition; (**b**) helicoidal conformation of PEO chains [95]. (Copyright (1992) American Chemical Society [95])

the interlayer space can resolve the racemic mixture of optically active metal complexes, and such properties can be employed in clay column chromatography for optical resolution [86].

(c) Electroconductive intercalation compounds with organic polymers [87–89]

The interactions between clay and organic polymers have been investigated over a long period of time, and various intercalation compounds have been prepared [90–94]. Their synthesis is mainly achieved by one of two different methods: direct intercalation of polymers form their solutions or their melts and intercalation of monomers and their subsequent polymerization in the interlayer space. One of the targets is electroconductive organic-inorganic hybrids. A typical example is the intercalation of poly(ethylene oxide), PEO, into a host substance bearing interlayer monovalent cations, such as Li^+ and Na^+. The intercalated PEO solvates interlayer cations and facilitates the transportation of Li^+ and Na^+, leading to ionic conductivity. The proposed structures of PEO-Li^+-montmorillonite is shown in Fig. 10.13. This strategy is attractive, since crystallization of PEO chains, occurring below 60 °C and reducing Li^+ ion conductivity significantly, can be suppressed in the interlayer space [88, 95, 96]. Another well-known example is oxidative polymerization of aniline in the interlayer space of layered substances containing appropriate transition metals. The resulting intercalation compounds exhibit electroconductivity [97–100].

(d) Preparation of nanosheets by exfoliation and their applications for polymer-based organic-inorganic hybrids [92, 93, 101–106]

Nanosheets can be prepared from layered materials via exfoliation. One of the common approaches to layered materials with negatively charged layers is intercalation of bulky ammonium ions, such as tetrabutylammonium ions, and subsequent treatment, such as ultrasonication and stirring, and the resulting nanosheets can be dispersed in water. The use of *n*-alkylammonium ions with long alkyl chains also causes exfoliation in organic solvents.

Grafting reactions can be employed for both neutral and negatively charged layered materials. In the case of layered materials with negatively charged layers, protonated forms can be employed. In addition, intercalation compounds with polar organic solvents and organic ions can be employed as intermediates. Since covalent bonds are formed via grafting reactions, interlayer surface modification by grafting reaction is appropriate for exfoliation. The introduction of polymer chains can be achieved by several techniques, including the surface-initiated atom transfer radical polymerization (SI-ATRP) process [107].

Exfoliation of graphite requires another process for completion [108]. The typical process involves the oxidation of graphite to form graphite oxide and subsequent ultrasonication leading to the formation of graphene oxide, which can be reduced to form graphene or modified with organic compounds using functional groups, such as epoxy groups or carboxyl groups, on the surfaces or at the edges.

Exfoliated nanosheets can be dispersed in polymer matrices to form nanosheet/polymer hybrids. Pioneering work has been reported for clay-nylon systems (clay-nylon hybrids) [109]. Mechanical and gas barrier properties were dramatically improved by the addition of relatively small amount of clay nanosheets. A large number of studies have been reported for polymer-based hybrids with clay nanosheets and other types of metal oxide nanosheets. To date, considerable attention has been paid to graphene-nanosheet-containing polymer-based hybrids [110–112]. Different approaches have been applied to improve their affinity to polymer matrices: binding previously prepared polymer chains to graphene nanosheets (grafting to method); growth of polymer chains from surfaces, typically by the SI-ATRP technique (grafting from method); and grafting functional groups that react with monomers during polymerization for matrix formation.

The combination of nanosheets and polymers can provide different types of organic-inorganic hybrid materials. Organic-inorganic hybrid films with layered structures can be obtained by sequential adsorption of cationic polymers and negatively charged nanosheets [113, 114]. Another type of organic-inorganic hybrid material are hydrogels, which exhibit unique mechanical properties [115, 116].

10.4 Surface Modification of Nanoparticles and the Application of Modified Nanoparticles to Polymer-Based Organic-Inorganic Hybrids [117–120]

Ceramic nanoparticles, especially metal oxide nanoparticles, are attracting increasing attention. The origin of this interest includes possible modification of the band structures of semiconductors and superparamagnetism of magnetic nanoparticles, such as magnetite nanoparticles, by downsizing.

Surface modification of metal oxide nanoparticles has been developed for several purposes depending on the objectives, and two issues can generally be pointed out. One is the suppression of aggregation. Nanoparticles tend to aggregate, and it is generally accepted that breakdown of the resulting aggregates into individual nanoparticles is difficult. Thus, surface modification is required to suppress the aggregation of metal oxide nanoparticles. Another issue is tuning of the surface properties to obtain sufficient affinity to media or matrices. In any application of nanoparticles, affinity to the surrounding environment is a key factor. Thus, an appropriate interface should be designed via surface modification.

(a) Surface modification process

Surface modification techniques were developed for modification of silica, and silylation has been applied [70, 71]. Thus, silylation reactions are also applied to other metal oxide nanoparticles. On the other hand, phosphorous coupling reagents are suitable for the modification of transition metal oxide nanoparticles and alumina nanoparticles, since M-O-P bonds formed are stable [121–123]. Phosphorous coupling reagents include organophosphonic acids ($RPO(OH)_2$) and phosphate ester (generally in the form of a mixture of monoester, $PO(OR)(OH)_2$, and diester, $PO(OR)_2(OH)$). POH groups can react with surface hydroxyls to form M-O-P bonds. Since phosphoryl groups act as Lewis base sites, moreover, their coordination to surface Lewis acid sites is possible. Thus, upon modification of an organophosphonic acid, for example, a bidentate or tridentate environment is expected to form on the surface, leading to stable modification (Fig. 10.14).

Carboxylic acids can also be employed for transition metal oxide nanoparticles. As in the case of the surface modification of organophosphonic acids, the formation of a bidentate environment is expected upon modification with a carboxylic acid.

(b) Application of surface-modified metal oxide nanoparticles [124–134]

Incorporation of metal oxide nanoparticles into organic polymers is currently one of the common strategies for improving polymer properties. Typical targets are high or low refractive index materials. By incorporating high refractive index nanoparticles, such as titania and zirconia nanoparticles [135, 136], or relatively low refractive index nanoparticles, such as silica [137], the refractive indices of organic-inorganic hybrid materials based on conventional polymers such as epoxy resins can be controlled.

Superparamagnetic nanoparticles are also attracting enormous attention because of their possible use in biomedical applications such as hyperthermia, drug delivery system, and magnetic resonance imaging [138, 139]. For these applications, surface modification is extremely important [139, 140]. Figure 10.15 shows surface-modified magnetic nanoparticles with multi functions [128]. Since magnetite is attractive for use in these applications, various approaches to surface modification of magnetite have been developed.

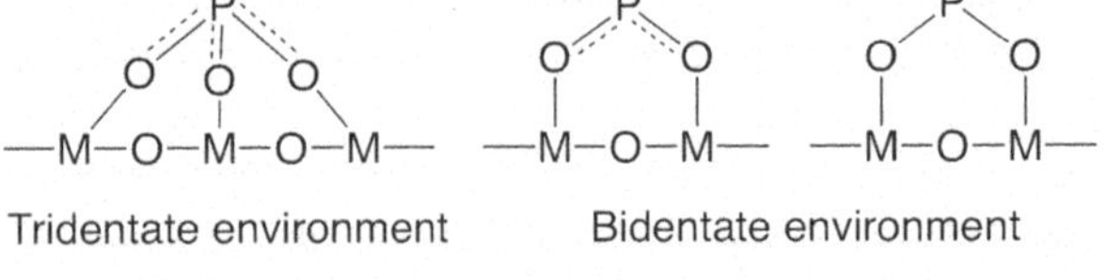

Fig. 10.14 Surface modification with an organophosphonic acid

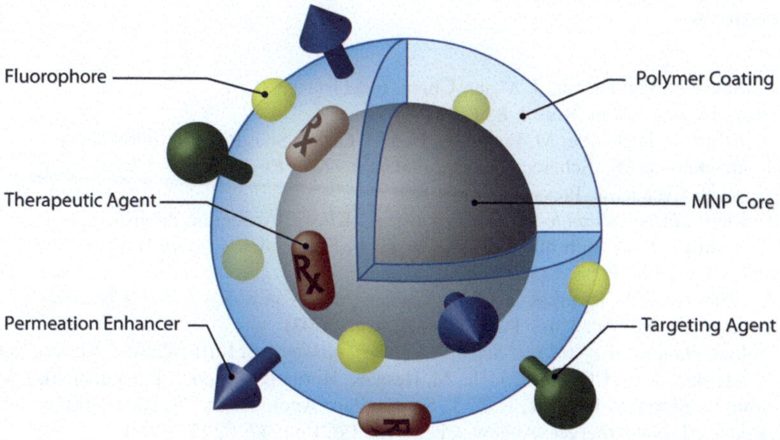

Fig. 10.15 Surface-modified magnetic nanoparticles (MNP) with multi functions. (Reprinted from Ref. [128], Copyright (2008), with permission from Elsevier [128])

Recently developed applications for sol-gel-derived mesoporous silica nanoparticles include their use in drug delivery systems [141, 142]. The advantages of mesoporous silica nanoparticles include their large and controllable pore volume, facile inner and outer surface modification, facile particle size control, and high biocompatibility. The introduction of molecular valves that can react to stimuli provides further control of the release of drugs.

10.5 Summary and Outlook

In this chapter, the preparation and some applications for organic-inorganic hybrid materials in which ceramic materials play key roles are described: sol-gel materials including mesoscaled materials, intercalation compounds and nanosheet-related materials, and organically modified nanoparticles. Preparation techniques have been developed for these materials and new types of materials have been revealed in recent decades. It is clear from society's demands for materials that the development of organic-inorganic hybrid materials is another major direction in ceramic material research, and one of the keys is the development of the precise preparation of inorganic component employed in organic-inorganic hybrid materials.

References

1. P. Judeinstein, C. Sanchez, J. Mater. Chem. **6**, 511 (1996)
2. E. Ruiz-Hitzky, Chem. Rec. **3**, 88 (2003)
3. C. Sanchez, P. Belleville, M. Popall, L. Nicole, Chem. Soc. Rev. **40**, 696 (2011)
4. C.J. Brinker, G.W. Scherer, *Sol-Gel Science. The Physics and Chemistry of Sol-Gel Processing* (Academic Press, Boston, 1990)
5. A.C. Pierre, *Introduction to sol-Gel Processing* (Kluwer Academic Publishers, Boston, 1998)
6. D.C. Bradley, R.C. Mehrotra, I.P. Rothwell, A. Singh, *Alkoxo and Aryloxo Derivatives of Metals* (Academic Press, London, 2001)
7. N.Y. Turova, E.P. Turevskaya, V.G. Kessler, M.I. Yanovskaya, *The Chemistry of Metal Alkoxides* (Kluwer Academic Publishers, Boston, 2002)
8. S. Sakka, *Handbook of sol-Gel Science and Technology*, vol I–III (Kluwer, Boston, 2005)
9. J. V. Alemán, A. V. Chadwick, J. He, M. Hess, K. Horie, R. G. Jones, P. Kratochvíl, I. Meisel, I. Mita, G. Moad, S. Penczek, R. F. T. Stepto, Pure Appl. Chem. **79**, 1801 (2007)
10. N. Pinna, M. Niederberger, Angew. Chem. Int. Ed. Eng. **47**, 5292 (2008)
11. P.H. Mutin, A. Vioux, Chem. Mater. **21**, 582 (2009)
12. J. Livage, C. Sanchez, J. Non-Cryst. Solids **145**, 11 (1992)
13. J.N. Hay, H.M. Raval, Chem. Mater. **13**, 3396 (2001)
14. R.C. Mehrotra, Struct. Bond. **77**, 1 (1992)
15. J.D. Mackenzie, E.P. Bescher, Acc. Chem. Res. **40**, 810 (2007)
16. H. Schmidt, ACS Symp. Ser. **360**, 333 (1988)
17. J. Wen, G.L. Wilkes, Chem. Mater. **8**, 1667 (1996)
18. D.A. Loy, K.J. Shea, Chem. Rev. **95**, 1431 (1995)
19. Y. Chujo, E. Ihara, S. Kure, T. Saegusa, Macromolecules **26**, 5681 (1993)
20. B.M. Novak, Adv. Mater. **5**, 422 (1993)
21. R.J.P. Corriu, Angew. Chem. Int. Ed. **39**, 1376 (2000)
22. K.J. Shea, D.A. Loy, Chem. Mater. **13**, 3306 (2001)
23. K.J. Shea, D.A. Loy, Acc. Chem. Res. **34**, 707 (2001)
24. B. Boury, R. Corriu, Chem. Rec. **3**, 120 (2003)
25. L.-C. Hu, K.J. Shea, Chem. Soc. Rev. **40**, 688 (2011)
26. P.F.W. Simon, R. Ulrich, H.W. Spiess, U. Wiesner, Chem. Mater. **13**, 3464 (2001)
27. M.E. Davis, Nature **417**, 813 (2002)
28. F. Hoffmann, M. Cornelius, J. Morell, M. Fröba, Angew. Chem. Int. Ed. **45**, 3216 (2006)
29. N. Mizoshita, T. Tani, S. Inagaki, Chem. Soc. Rev. **40**, 789 (2011)
30. R. Muñoz-Espí, C.K. Weiss, K. Landfester, Curr. Opin. Colloid Interface Sci. **17**, 212 (2012)
31. K. Ariga, Chem. Rec. **3**, 297 (2004)
32. J.J.E. Moreau, L. Vellutini, M.W.C. Man, C. Bied, J. Am. Chem. Soc. **123**, 1509 (2001)
33. J.J.E. Moreau, L. Vellutini, M.W.C. Man, C. Bied, J.L. Bantignies, P. Dieudonné, J.L. Sauvajol, J. Am. Chem. Soc. **123**, 7957 (2001)
34. N.K. Raman, M.T. Anderson, C.J. Brinker, Chem. Mater. **8**, 1682 (1996)
35. J.Y. Ying, C.P. Mehnert, M.S. Wong, Angew. Chem. Int. Ed. **38**, 56 (1999)
36. F. Schüth, Chem. Mater. **13**, 3184 (2001)
37. C.T. Kresge, M.E. Leonowicz, W.J. Roth, J.C. Vartuli, J.S. Beck, Nature **359**, 710 (1992)
38. J.S. Beck, J.C. Vartuli, W.J. Roth, M.E. Leonowicz, C.T. Kresge, K.D. Schmitt, C.T.W. Chu, D.H. Olson, E.W. Sheppard, S.B. McCullen, J.B. Higgins, J.L. Schlenker, J. Am. Chem. Soc. **114**, 10834 (1992)
39. A. Shimojima, Y. Sugahara, K. Kuroda, J. Am. Chem. Soc. **120**, 4528 (1998)
40. A. Shimojima, K. Kuroda, Angew. Chem. Int. Ed. **42**, 4057 (2003)
41. S. Inagaki, S. Guan, T. Ohsuna, O. Terasaki, Nature **416**, 304 (2002)
42. G.E. Fryxell, Inorg. Chem. Commun. **9**, 1141 (2006)
43. C. Sanchez, B. Julián, P. Belleville, M. Popall, J. Mater. Chem. **15**, 3559 (2005)

44. P.N. Prasad, F.V. Bright, U. Narang, R. Wang, R.A. Dunbar, J.D. Jordan, R. Gvishi, ACS Symp. Ser. **585**, 317 (1995)
45. C. Sanchez, F. Ribot, B. Lebeau, J. Mater. Chem. **9**, 35 (1999)
46. B. Lebeau, C. Sanchez, Curr. Opin. Solid State Mater. Sci. **4**, 11 (1999)
47. C. Sanchez, B. Lebeau, F. Chaput, J.P. Boilot, Adv. Mater. **15**, 1969 (2003)
48. B. Dunn, J.I. Zink, Chem. Mater. **9**, 2280 (1997)
49. J. Biteau, F. Chaput, J.-P. Boilot, J. Phys. Chem. **100**, 9024 (1996)
50. B. Lebeau, S. Brasselet, J. Zyss, C. Sanchez, Chem. Mater. **9**, 1012 (1997)
51. K. Moller, T. Bein, Chem. Mater. **10**, 2950 (1998)
52. B.J. Scott, G. Wirnsberger, G.D. Stucky, Chem. Mater. **13**, 3140 (2001)
53. G. Schulz-Ekloff, D. Wöhrle, B. van Duffel, R.A. Schoonheydt, Microporous Mesoporous Mater. **51**, 91 (2002)
54. J. Feng, H. Zhang, Chem. Soc. Rev. **42**, 387 (2013)
55. I. Gill, A. Ballesteros, Trends Biotechnol. **18**, 282 (2000)
56. J. Livage, T. Coradin, C. Roux, J. Phys. Condens. Matter **13**, R673 (2001)
57. I. Gill, Chem. Mater. **13**, 3404 (2001)
58. A.C. Pierre, Biocatal. Biotransform. **22**, 145 (2004)
59. G.J. Owens, R.K. Singh, F. Foroutan, M. Alqaysi, C.-M. Han, C. Mahapatra, H.-W. Kim, J.C. Knowles, Prog. Mater. Sci. **77**, 1 (2016)
60. H. Frenkel-Mullerad, D. Avnir, J. Am. Chem. Soc. **127**, 8077 (2005)
61. B.K.G. Theng, *The Chemistry of Clay-Organic Reactions* (Adam Hilger, London, 1974)
62. M.S. Whittingham, A.J. Jacobson, *Intercalatioin Chemistry* (Academic Press, New York, 1982)
63. G. Lagaly, Solid State Ionics **22**, 43 (1986)
64. G. Lagaly, K. Beneke, Colloid Polym. Sci. **269**, 1198 (1991)
65. A.J. Jacobson, in *Solid State Chemistry: Compounds*, ed. by A. K. Cheetham, P. Day, (Oxford University Press, Oxford, 1992), p. 182
66. P. O'Brien, in *Inorganic Materials, 2nd edn.* ed. by D. W. Bruce, D. O'Hare, (John Wiley & Sons, Chichester, 1997), p. 171
67. M. Ogawa, K. Kuroda, Bull. Chem. Soc. Jpn. **70**, 2593 (1997)
68. E. Ruiz-Hitzky, J.M. Rojo, Nature **287**, 28 (1980)
69. N. Takahashi, K. Kuroda, J. Mater. Chem. **21**, 14336 (2011)
70. E.P. Plueddemann, *Silane Coupling Agents*, 2nd edn. (Springer Science + Business Media, New York, 1991)
71. P. Van Der Voort, E.F. Vansant, J. Liq. Chromatogr. Relat. Technol. **19**, 2723 (1996)
72. Y. Ide, M. Ogawa, Angew. Chem. Int. Ed. **46**, 8449 (2007)
73. N. Kimura, Y. Kato, R. Suzuki, A. Shimada, S. Tahara, T. Nakato, K. Matsukawa, P.H. Mutin, Y. Sugahara, Langmuir **30**, 1169 (2014)
74. A.B. Bourlinos, D. Gournis, D. Petridis, T. Szabó, A. Szeri, I. Dékány, Langmuir **19**, 6050 (2003)
75. M. Ogawa, K. Kuroda, Chem. Rev. **95**, 399 (1995)
76. T.J. Pinnavaia, Science, **220**, 365 (1983)
77. R. Sasai, N. Iyi, T. Fujita, F.L. Arbeloa, K. Takagi, H. Itoh, Langmuir **20**, 4715 (2004)
78. M. Ogawa, M. Takahashi, K. Kuroda, Chem. Mater. **6**, 715 (1994)
79. R. Takenawa, Y. Komori, S. Hayashi, J. Kawamata, K. Kuroda, Chem. Mater. **13**, 3741 (2001)
80. J. Tudor, L. Willington, D. O'Hare, B. Royan, Chem. Commun., 2031 (1996)
81. T. Baskaran, J. Christopher, A. Sakthivel, RSC Adv. **5**, 98853 (2015)
82. C.V. Kumar, A. Chaudhari, J. Am. Chem. Soc. **122**, 830 (2000)
83. A. Yamagishi, J. Coord. Chem. **16**, 131 (1987)
84. M. Ogawa, S. Okutomo, K. Kuroda, J. Am. Chem. Soc. **120**, 7361 (1998)
85. J.W. Johnson, A.J. Jacobson, W.M. Butler, S.E. Rosenthal, J.F. Brody, J.T. Lewandowski, J. Am. Chem. Soc. **111**, 381 (1989)
86. A. Yamagishi, M. Taniguchi, Y. Imamura, H. Sato, Appl. Clay Sci. **11**, 1 (1996)
87. E. Ruiz-Hitzky, Adv. Mater. **5**, 334 (1993)

88. E. Ruiz-Hitzky, P. Aranda, B. Casal, J.C. Galván, Adv. Mater. **7**, 180 (1995)
89. E. Ruiz-Hitzky, B. Casal, P. Aranda, J.C. Galván, Rev. Inorg. Chem. **21**, 125 (2001)
90. B.K.G. Theng, *Formation and Properties of Clay-Polymer Complexes, 2nd edn.* (Elsevier, Amsterdam, 2012)
91. E.P. Giannelis, Appl. Organomet. Chem. **12**, 675 (1998)
92. M. Alexandre, P. Dubois, Mater. Sci. Eng. R **28**, 1 (2000)
93. S.S. Ray, M. Okamoto, Prog. Polym. Sci. **28**, 1539 (2003)
94. S. Pavlidou, C.D. Papaspyrides, Prog. Polym. Sci. **33**, 1119 (2008)
95. P. Aranda, E. Ruiz-Hitzky, Chem. Mater. **4**, 1395 (1992)
96. R.A. Vaia, S. Vasudevan, W. Krawiec, L.G. Scanlon, E.P. Giannelis, Adv. Mater. **7**, 154 (1995)
97. M.G. Kanatzidis, C.G. Wu, H.O. Marcy, C.R. Kannewurf, J. Am. Chem. Soc. **111**, 4139 (1989)
98. M.G. Kanatzidis, C.-G. Wu, H.O. Marcy, D.C. DeGroot, C.R. Kannewurf, A. Kostikas, V. Papaefthymiou, Adv. Mater. **2**, 364 (1990)
99. M.G. Kanatzidis, R. Bissessur, D.C. Degroot, J.L. Schindler, C.R. Kannewurf, Chem. Mater. **5**, 595 (1993)
100. L. Wang, P. Brazis, M. Rocci, C.R. Kannewurf, M.G. Kanatzidis, Chem. Mater. **10**, 3298 (1998)
101. A. Okada, A. Usuki, Mater. Sci. Eng. C **3**, 109 (1995)
102. C. Tan, H. Zhang, Chem. Soc. Rev. **44**, 2713 (2015)
103. V. Nicolosi, M. Chhowalla, M.G. Kanatzidis, M.S. Strano, J.N. Coleman, Science **340**, 1226419 (2013)
104. J.N. Coleman, M. Lotya, A. O'neill, S.D. Bergin, P.J. King, U. Khan, K. Young, A. Gaucher, S. De, R.J. Smith, I.V. Shvets, S.K. Arora, G. Stanton, H.-Y. Kim, K. Lee, G.T. Kim, G.S. Duesberg, T. Hallam, J.J. Boland, J.J. Wang, J.F. Donegan, J.C. Grunlan, G. Moriarty, A. Shmeliov, R.J. Nicholls, J.M. Perkins, E.M. Grieveson, K. Theuwissen, D.W. McComb, P.D. Nellist, V. Nicolosi, Science **331**, 568 (2011)
105. R. Ma, T. Sasaki, Adv. Mater. **22**, 5082 (2010)
106. M. Osada, T. Sasaki, Adv. Mater. **24**, 210 (2012)
107. N. Idota, S. Fukuda, T. Tsukahara, Y. Sugahara, Chem. Lett. **44**, 203 (2015)
108. S. Mao, H. Pu, J. Chen, RSC Adv. **2**, 2643 (2012)
109. A. Usuki, Y. Kojima, M. Kawasumi, A. Okada, Y. Fukushima, T. Kurauchi, O. Kamigaito, J. Mater. Res. **8**, 1179 (1993)
110. S. Stankovich, D.A. Dikin, G.H.B. Dommett, K.M. Kohlhaas, E.J. Zimney, E.A. Stach, R.D. Piner, S.T. Nguyen, R.S. Ruoff, Nature **442**, 282 (2006)
111. T. Kuilla, S. Bhadra, D. Yao, N.H. Kim, S. Bose, J.H. Lee, Prog. Polym. Sci. **35**, 1350 (2010)
112. T. Kuila, S. Bose, A.K. Mishra, P. Khanra, N.H. Kim, J.H. Lee, Prog. Mater. Sci. **57**, 1061 (2012)
113. E.R. Kleinfeld, G.S. Ferguson, Science **265**, 370 (1994)
114. S.W. Keller, H.-N. Kim, T.E. Mallouk, J. Am. Chem. Soc. **116**, 8817 (1994)
115. S. Tamesue, M. Ohtani, K. Yamada, Y. Ishida, J.M. Spruell, N.A. Lynd, C.J. Hawker, T. Aida, J. Am. Chem. Soc. **135**, 15650 (2013)
116. M. Liu, Y. Ishida, Y. Ebina, T. Sasaki, T. Hikima, M. Takata, T. Aida, Nature **517**, 68 (2015)
117. F. Caruso, Adv. Mater. **13**, 11 (2001)
118. M.Z. Rong, M.Q. Zhang, W.H. Ruan, Mater. Sci. Technol. **22**, 787 (2006)
119. M.-A. Neouze, U. Schubert, Monatsh. Chem. **139**, 183 (2008)
120. S. Kango, S. Kalia, A. Celli, J. Njuguna, Y. Habibi, R. Kumar, Prog. Polym. Sci. **38**, 1232 (2013)
121. C. Queffélec, M. Petit, P. Janvier, D.A. Knight, B. Bujoli, Chem. Rev. **112**, 3777 (2012)
122. G. Guerrero, J.G. Alauzun, M. Granier, D. Laurencin, P.H. Mutin, Dalton Trans. **42**, 12569 (2013)
123. R. Boissezon, J. Muller, V. Beaugeard, S. Monge, J.-J. Robin, RSC Adv. **4**, 35690 (2014)

124. W. Caseri, Macromol. Rapid Commun. **21**, 705 (2000)
125. R.J. Nussbaumer, W.R. Caseri, P. Smith, T. Tervoort, Macromol. Mater. Eng. **288**, 44 (2003)
126. A.K. Gupta, M. Gupta, Biomaterials **26**, 3995 (2005)
127. I.I. Slowing, J.L. Vivero-Escoto, C.-W. Wu, V.S.-Y. Lin, Adv. Drug Deliv. Rev. **60**, 1278 (2008)
128. C. Sun, J.S.H. Lee, M. Zhang, Adv. Drug Deliv. Rev. **60**, 1252 (2008)
129. K.K. Coti, M.E. Belowich, M. Liong, M.W. Ambrogio, Y.A. Lau, H.A. Khatib, J.I. Zink, N.M. Khashab, J.F. Stoddart, Nanoscale **1**, 16 (2009)
130. J.H. Jang, H.B. Lim, Microchem. J. **94**, 148 (2010)
131. R.A. Sperling, W.J. Parak, Philos. Trans. R. Soc. A **368**, 1333 (2010)
132. S. Laurent, S. Dutz, U.O. Häfeli, M. Mahmoudi, Adv. Colloid Interf. Sci. **166**, 8 (2011)
133. N. Lee, D. Yoo, D. Ling, M.H. Cho, T. Hyeon, J. Cheon, Chem. Rev. **115**, 10637 (2015)
134. A. Watermann, J. Brieger, Nanomater. **7**, 189 (2017)
135. W.-C. Chen, S.-J. Lee, L.-H. Lee, J.-L. Lin, J. Mater. Chem. **9**, 2999 (1999)
136. K. Luo, S. Zhou, L. Wu, Thin Solid Films **517**, 5974 (2009)
137. K.C Krogman, T. Druffel, M.K Sunkara, Nanotechnology **16**, S338 (2005)
138. A. Jordan, R. Scholz, P. Wust, H. Fähling, R. Felix, J. Magn. Magn. Mater. **201**, 413 (1999)
139. C. Wilhelm, C. Billotey, J. Roger, J.N. Pons, J.C. Bacri, F. Gazeau, Biomaterials **24**, 1001 (2003)
140. Y. Zhang, N. Kohler, M. Zhang, Biomaterials **23**, 1553 (2002)
141. J. Lu, M. Liong, J.I. Zink, F. Tamanoi, Small **3**, 1341 (2007)
142. Y. Zhao, B.G. Trewyn, I.I. Slowing, V.S.-Y. Lin, J. Am. Chem. Soc. **131**, 8398 (2009)

The manufacturer's authorised representative in the EU is Springer
Nature Customer Service Centre GmbH, Europaplatz 3, 69115 Heidelberg,
Germany. If you have any concerns regarding our products, please
contact ProductSafety@springernature.com

Printed and bound by CPI Group (UK) Ltd, Croydon, CR0 4YY
30/09/2026
02225556-0001